Strongly Interacting Quantum Systems, Volume 1

Few-body physics

Online at: https://doi.org/10.1088/978-0-7503-3087-9

IOP Series in Quantum Technology

About the series

The IOP Series in Quantum Technology is dedicated to bringing together the most up to date texts and reference books from across the emerging field of quantum science and its technological applications. Prepared by leading experts, the series is intended for graduate students and researchers either already working in or intending to enter the field. The series seeks (but is not restricted to) publications in the following topics:

- Quantum biology
- Quantum communication
- Quantum computation
- Quantum control
- Quantum cryptography
- Quantum engineering
- Quantum machine learning and intelligence
- Quantum materials
- Quantum metrology
- Quantum optics
- Quantum sensing
- Quantum simulation
- Quantum software, algorithms and code
- Quantum thermodynamics
- Hybrid quantum systems

A full list of titles published in this series can be found here: https://iopscience.iop.org/bookListInfo/iop-series-in-quantum-technology.

Strongly Interacting Quantum Systems, Volume 1

Few-body physics

Manuel Valiente
University of Murcia, C. Campus Universitario, 11, 30100 Murcia, Spain

Nikolaj Thomas Zinner
Aarhus University, Aarhus, Denmark

IOP Publishing, Bristol, UK

ISBN 978-0-7503-3087-9 (ebook)
ISBN 978-0-7503-3085-5 (print)
ISBN 978-0-7503-3088-6 (myPrint)
ISBN 978-0-7503-3086-2 (mobi)

DOI 10.1088/978-0-7503-3087-9

Version: 20230901

IOP ebooks

British Library Cataloguing-in-Publication Data: A catalogue record for this book is available from the British Library.

Published by IOP Publishing, wholly owned by The Institute of Physics, London

IOP Publishing, No.2 The Distillery, Glassfields, Avon Street, Bristol, BS2 0GR, UK

US Office: IOP Publishing, Inc., 190 North Independence Mall West, Suite 601, Philadelphia, PA 19106, USA

A mis niños: Manuel K, David H, Antonio y Álvaro.

—M Valiente

Til Tine, Krümmel og Chloé

—N T Zinner

Contents

Preface

Quantum few-body physics is a very extensive field dedicated to the study of microscopic objects, such as elementary particles, atoms, nuclei, or molecules, whose constituents or ensembles can be described with a few degrees of freedom. There is a huge variety of methods and theories used to approach these systems, ranging from relativistic quantum field theory, non-relativistic quantum mechanics, to semi-classical theories. Few-body systems are interesting and important on their own. Many of them are experimentally studied in both large and small facilities, especially colliders, where scattering cross sections are routinely measured for a number of ever-growing few-body processes. Collision experiments are the main tools used to study many important phenomena, among many others, such as spectroscopic measurements of atoms, nuclei and molecules. However, during the last 15 years, advances in trapping and manipulating neutral atoms that are cooled to near-zero temperatures have resulted in a number of spectacular few-body experimental observations of phenomena that had been elusive in other systems for a very long time. A prime example of this is the Efimov effect, predicted to occur in 1970, with its first experimental evidence confirmed in a groundbreaking experiment with ultracold caesium in 2006 by R Grimm and coworkers at the University of Innsbruck. Ultracold atoms have since become a very interesting platform on which quantum few-body physics can be studied, and much of this book is motivated by advances in experimental as well as theoretical ultracold atomic physics.

From a theorist's point of view, few-body systems have a major advantage when compared with their many-body counterparts. Namely, the lower complexity of having one, two or three particles, results in the possibility of studying certain properties of few-body systems *very* accurately, and sometimes even exactly. This means that in many situations we may be able to deal with highly non-perturbative phenomena, and access important properties of strongly-interacting quantum systems. The small number of degrees of freedom, however, does not mean few-body physics is in general not complicated, but is in fact the opposite. We need to simplify interactions as much as possible to be able to make physically sound predictions about the system of interest *in the energy range of interest*.

The purpose of this volume is to introduce the reader to the physics of few, strongly interacting quantum particles. Instead of trying to review the extensive literature on the subject, we focus on non-relativistic few-body systems both in free space and in structured media, such as crystalline lattices, optical lattices and external traps. To study strong interactions, we put strong emphasis on low-energy phenomenology, which requires heavy use of regularization and renormalization techniques even in the non-relativistic limit. These methods are explained thoroughly, and are illustrated using models of current interest in atomic, molecular, optical, solid state and nuclear physics alike. This textbook is aimed at advanced MSc and PhD students, and we assume the reader is familiar with quantum mechanics as well as the basics of non-relativistic quantum scattering theory. Most derivations in the book are detailed, with the purpose of allowing the interested reader to study this volume on their own. Wherever possible, we have also tried to give several different methods to solve one problem, and not only the easiest or most difficult one. We expect the book will be useful as a reference to pursue further developments in the field to experts and young scientists alike.

Writing this book has not been an easy task. We wrote during a global pandemic, got stranded out-of-office, and went through several faculty changes. We thank our Editor at IOP Publishing, John Navas, for his immense patience every time our book was delayed. We could not have started or finished this project without the support of many colleagues with whom we have discussed and collaborated in this and other topics, and the funding agencies that have supported our work. We would like to thank our colleagues C W Duncan, P Öhberg, L G Phillips, H Zhai, V Pastukhov, A G Volosniev, O V Marchukov, D V Fedorov, A S Jensen, J R Armstrong, H-W Hammer, K Mølmer, D Petrosyan, A Saenz, K Fraser, B Juliá-Díaz, I Morera, S Jerez, M A García-March, S Mukherjee, E Andersson, M J Edmonds, L Santos, Y Nishida, Y Sekino, F F Bellotti, T Frederico, M T Yamashita, L Tomio, J B Hofmann and X Chen. Special thanks go to Xiaoling Cui, who, much to our regret, was occupied with far more important ventures and could not partake in the writing of this book. We hope she will enjoy these pages, many of which are inspired by her work. We acknowledge support from the Ramón y Cajal programme, Spanish Ministry of Science and Innovation through grant number RYC2020-029961-I, as well as the national research and development grant PID2021-126039NA-I00.

Manuel Valiente
Tenerife, Spain
Nikolaj Thomas Zinner
Aarhus, Denmark

Author biographies

Manuel Valiente

Manuel Valiente is currently a Ramón y Cajal Fellow at the University of Murcia, where he moved after one year at the University of La Laguna. He was a postdoctoral researcher at several institutions around the world, and previously held a Faculty position at Tsinghua University in Beijing. He obtained a PhD in Physics at Humboldt-Universität zu Berlin in Germany in 2010.

Nikolaj Thomas Zinner

Nikolaj Thomas Zinner is Professor of theoretical physics at the Department of Physics at Aarhus University, Denmark, where he works mainly in quantum dynamics and quantum technology. He was a postdoctoral fellow at Harvard University and at the Niels Bohr Institute before becoming a faculty member at Aarhus University. Nikolaj obtained his PhD in theoretical nuclear physics in 2007 from Aarhus University.

IOP Publishing

Strongly Interacting Quantum Systems, Volume 1
Few-body physics
Manuel Valiente and Nikolaj Thomas Zinner

Chapter 1

Fundamentals of quantum mechanics and scattering theory

In this chapter, we begin by introducing some basic concepts in quantum mechanics, mostly with the purpose of setting the notation and definitions. We then build the standard non-relativistic scattering theory, which is very important throughout the rest of the book.

1.1 Quantum mechanics

To describe microscopic systems accurately, we are required to use quantum mechanics. The theory gives us a framework with which we can study strongly interacting particles, whether it be atoms or electrons, for instance, fully non-perturbatively. In this section we introduce some important concepts that are assumed tp be known to the reader but can be useful both as a reference and for notational purposes.

1.1.1 Hilbert spaces

A quantum-mechanical state is represented by a vector in a Hilbert space. The Hilbert space depends on the particular system of interest. For instance, a non-relativistic single particle in one spatial dimension under the influence of a quadratic potential (the quantum harmonic oscillator) has an associated Hilbert space within the $L^2(\mathbb{R})$ space of square-integrable functions. The quantum states of an N-body system in vacuum, in D spatial dimensions, however, lie in $L^2(\mathbb{R}^{ND})$, while the state of a particle on a one-dimensional lattice is represented by a square-summable function in $\ell^2(\mathbb{Z})$. The scalar products for one-dimensional continuous and discrete spaces are given by

$$\langle\phi|\psi\rangle = \int dx \phi^*(x)\psi(x), \tag{1.1}$$

doi:10.1088/978-0-7503-3087-9ch1

for L^2-functions, and by

$$\langle \phi | \psi \rangle = \sum_{n=-\infty}^{\infty} \phi^*(n)\psi(n), \tag{1.2}$$

for ℓ^2-functions. As usual, the norm of a vector in a Hilbert space is the one derived from its scalar product, namely

$$||\psi||^2 = \langle \psi | \psi \rangle. \tag{1.3}$$

A state in a Hilbert space will be typically written in terms of an orthonormal basis. When this is discrete, labelled with an index n, we will write

$$\psi(x) = \sum_{n=0}^{\infty} c_n \phi_n(x), \tag{1.4}$$

where $\langle \phi_{n'} | \phi_n \rangle = \delta_{n,n'}$. We may also use, in L^2, a complete, orthonormal set of continuum states, in which case

$$\psi(x) = \int_{-\infty}^{\infty} ds c(s) \phi_s(x), \tag{1.5}$$

where $\langle \phi_{s'} | \phi_s \rangle = \delta(s - s')$. The most general situation will involve a sum over discrete states and an integral over continuum states of a complete basis. In many cases, such as a finite pocket trapping particles, the sum over discrete states may be a finite sum. Of course, for the important case of Coulomb attraction, there is an infinite mumber of discrete states.

1.1.2 Schrödinger equation

A quantum-mechanical state will evolve over time according to the Schrödinger equation. For a general system with quantum Hamiltonian $H(t)$, the state will obey

$$H(t)|\psi(t)\rangle = i\hbar\partial_t|\psi(t)\rangle. \tag{1.6}$$

If the Hamiltonian is time independent, then the Schrödinger equation (1.6) can be reduced to an eigenvalue problem, called the stationary Schrödinger equation. To obtain this, we define $|\psi(t)\rangle = |\psi\rangle f(t)$, where all time dependence is encoded in the function $f(t)$. The Schrödinger equation separates as $f(t) = \exp(-iEt/\hbar)$ and

$$H|\psi\rangle = E|\psi\rangle. \tag{1.7}$$

The Hamiltonian must be a self-adjoint operator, and therefore have a real spectrum, that is, the set of energies E which make equation (1.7) have L^2-solutions, which comprise the discrete states and eigenvalues, as well as those energies that comprise the continuous spectrum and allow for generalized eigenstates that are bounded as $|x| \to \infty$ but do not properly belong to L^2. The former typically describes quantum-mechanical bound states, while the latter are scattering states.

Our first goal when we deal with any problem of interacting quantum particles will be to solve the Schrödinger equation or, at least, obtain some of its solutions at the energies of interest. Mostly, we will deal with time-independent Hamiltonians in this book and, consequently, much of the material is devoted to solving the stationary Schrödinger equation.

1.1.3 Position and momentum representations

There are two sets of fundamental operators in quantum theory that deserve special treatment, namely the position and momentum operators. In D spatial dimensions, these are given by sets of D operators, one for each Cartesian component. In three dimensions, for instance, one would define $\mathbf{R} = (X,\ Y,\ Z)$ for the position operators, and call their eigenstates, respectively $|x\rangle$, $|y\rangle$ and $|z\rangle$, and their eigenvalues x, y and z. We may, therefore, write,

$$X|x\rangle = x|x\rangle, \tag{1.8}$$

and analogously for the remaining components. Since the position operators are physical observables, they are self-adjoint, and therefore x is real, and $\langle x|X = \langle x|x$. Self-adjointness also implies that the set $\{|x\rangle\}_{x\in\mathbb{R}}$ is orthonormal and, since the spectrum of X is by definition continuous, orthonormality takes the form

$$\langle x'|x\rangle = \delta(x - x'). \tag{1.9}$$

The position eigenbasis is also complete, meaning that any state in the Hilbert space $L^2(\mathbb{R})$ can be written as a linear combination of their members. Formally, if $|\psi\rangle$ is in the Hilbert space, then

$$|\psi\rangle = \int_{-\infty}^{\infty} dx\psi(x)|x\rangle, \tag{1.10}$$

where $\psi(x)$ are the expansion coefficients in the *position representation.* By taking the scalar product with some position eigenstate, say $|x\rangle$, we obtain from equation (1.10)

$$\langle x|\psi\rangle = \psi(x), \tag{1.11}$$

where we have used the orthonormality condition (1.9). We now obtain the resolution of the identity in the position representation. To see this, we claim

$$1 = \int dx|x\rangle\langle x|. \tag{1.12}$$

We apply the operator on the right-hand-side of equation (1.12) on an arbitrary vector $|\psi\rangle$ in the Hilbert space, obtaining

$$\left[\int dx|x\rangle\langle x|\right]|\psi\rangle = \int dx|x\rangle\langle x|\psi\rangle = \int dx|x\rangle\psi(x). \tag{1.13}$$

equation (1.13) is identical to equation (1.10), which means that the application of the right-hand-side of equation (1.12) on *any* vector gives exactly that vector, that is, it is the identity operator. Hence, we have proved equation (1.12).

We now introduce the momentum operators, which are denoted $\mathbf{P} = (P_x, P_y, P_z)$ in three dimensions, with obvious notation. By definition, P_x is conjugated to X, which is to say that, by definition,

$$[X, P_x] = i\hbar. \tag{1.14}$$

We write the eigenstates of P_x associated with the eigenvalue p_x as $|p_x\rangle$. The orthogonality condition we shall use is slightly different than that for position eigenstates which by simple convention is usual to scattering theory. We shall write it as

$$\langle p_x'|p_x\rangle = \hbar^2(2\pi)\delta(p_x - p_x'). \tag{1.15}$$

The above relation looks cumbersome, but we will see shortly and throughout the book that it is very convenient. Since P_x is self-adjoint, and has a continuous spectrum being the whole real line, we can write any vector $|\psi\rangle$ in the Hilbert space as

$$|\psi\rangle = \int dp_x \psi(p_x)|p_x\rangle. \tag{1.16}$$

To see what the momentum operator looks like in the position representation, from the formalism itself, we apply it to equation (1.16). We have

$$\langle x|P_x|\psi\rangle = \int dp_x p_x \psi(p_x)\langle x|p_x\rangle. \tag{1.17}$$

If ψ is itself an (generalized) eigenstate of P_x with (generalized) eigenvalue p_x', we have

$$p_x'\psi(x) = \int dp_x p_x \psi(p_x)\langle x|p_x\rangle. \tag{1.18}$$

The solution requires $\psi(p_x) \propto \delta(p_x - p_x')$ and

$$\psi(x) = \langle x|p_x'\rangle. \tag{1.19}$$

Therefore, we have

$$\psi(p_x) \propto \delta(p_x - p_x') \propto \int dx e^{i(p_x - p_x')x/\hbar} \tag{1.20}$$

The position operator applied on $\psi(p_x)$, therefore, operates as

$$X\psi(p_x) \propto \int dx x e^{i(p_x - p_x')x/\hbar} = -i\hbar\frac{\partial}{\partial p_x}\int dx e^{i(p_x - p_x')/\hbar}, \tag{1.21}$$

which means that, acting on a general momentum-represented wave function, we have

$$X\psi(p_x) = -\frac{i}{\hbar}\frac{\partial}{\partial p_x}\psi(p_x). \tag{1.22}$$

We can carry out the reciprocal analysis for the momentum operator to find the well known relation, in the position representation, that

$$P_x\psi(x) = -i\hbar\frac{\partial}{\partial x}\psi(x). \tag{1.23}$$

Note that both representations are fully consistent with the canonical commutation relation (1.14).

As we already noted, the normalization condition in equation (1.15) looks cumbersome. However, there is a reason for that. Instead of using the linear momentum as a (dimensioned) label, we shall always use the wave number (or the wave vector) $k_x = p_x/\hbar$ ($\mathbf{k}$). In this way, the normalization is written as

$$\langle k_x'|k_x\rangle = 2\pi\delta(k_x - k_x'), \tag{1.24}$$

and the states themselves are given, in the position representation, via

$$\langle x|k_x\rangle = e^{ik_x x}. \tag{1.25}$$

Hence, the identity reads, on this basis,

$$1 = \frac{1}{2\pi}\int dk_x|k_x\rangle\langle k_x|. \tag{1.26}$$

This is the convention we follow throughout this book, which consists of writing $1/(2\pi)^D$ in front of every momentum integral in D space dimensions.

1.2 Non-relativistic scattering theory

In this section we describe, in a rather detailed fashion, the standard non-relativistic theory of quantum collisions. We take, for the most part, an operator approach, which allows us to consider scattering processes in non-standard systems such as electrons in crystals, which are not Galilean relativistic (i.e., not non-relativistic), yet not relativistic (i.e., do not require a quantum field theory). Other systems that can be described within the operator approach can indeed have a relativistic-like single-particle kinetic energy dispersion (e.g., $E(\mathbf{k}) = \pm\sqrt{(mc^2)^2 + \hbar|\mathbf{k}|^2c^2}$), but interactions that do not respect Lorentz invariance. These types of theories show up as low-energy effective descriptions of particles in structured media, a notable example of this being atoms in optical lattices with a particular geometry or electrons in graphene under certain conditions. The reader interested in the standard non-relativistic scattering theory is urged to read the excellent textbook of Taylor [1], which we partially follow.

In this section, unless otherwise stated, when studying non-relativistic scattering, we consider two particles in D spatial dimensions, of identical mass m, so that their reduced mass is $\mu = m/2$, and their relative kinetic energy is $\hbar^2k^2/m$ ($\mathbf{k} = |\mathbf{k}_1 - \mathbf{k}_2|/2$). We assume the centre of mass and relative coordinates have already been separated, so that we work with relative momentum.

1.2.1 The scattering operator (*S*-matrix)

The first thing we need in order to define a quantum theory of scattering is the actual definition of what scattering means. In a classical system, this is a simple matter: incoming particles (at asymptotically long past times) interact with each other at times around the present, and become essentially free at asymptotically long future times. That is, they move at constant speeds both long before and long after collision. What is interesting in this classical situation is the asymptotic future momenta given the asymptotic past momenta of the particles, while the intricate trajectory followed by the particles during the collision time is either not necessary or even unobservable. The analogue in quantum mechanics requires the definition of asymptotic states. Firstly, we consider a Hamiltonian H which can be split as $H = H_0 + H_I$, where H_0 contains all single-particle operators that support a continuous spectrum. In translationally invariant non-relativistic quantum mechanics, $H_0 = \mathbf{p}^2/m$ is the relative non-interacting Hamiltonian, after separation of centre of mass and relative motion. However, H_0 may consist of not only kinetic but also potential energy terms, such as periodic potentials or short-range one-body potentials, both supporting a continuous spectrum. The interaction term H_I is supposed to be short-ranged, in a way that supports scattering as we shall see shortly. The time-dependent Schrödinger equation with initial state $|\psi(t_0)\rangle$, equation (1.6), is solved as

$$|\psi(t)\rangle = U(t, t_0)|\psi(t_0)\rangle, \tag{1.27}$$

where $U(t, t_0)$ is the propagator. Since we will mainly work with time-independent Hamiltonians, the formal solution is given by

$$U(t, t_0) = e^{-iH(t-t_0)/\hbar}. \tag{1.28}$$

Naturally, we define the non-interacting, or free propagator $U_0(t, t_0)$ as that corresponding to free motion, that is

$$U_0(t, t_0) = e^{-iH_0(t-t_0)/\hbar}. \tag{1.29}$$

An asymptotic state $|\phi(t_0)\rangle$, just as in classical mechanics, should follow free motion. That is, $|\phi(t)\rangle = U_0(t, t_0)|\phi(t_0)\rangle$, for $t \to \pm\infty$. We prove now in dimension $D = 3$ that, for local interactions $\langle \mathbf{r}'|H_I|\mathbf{r}\rangle = \delta(\mathbf{r} - \mathbf{r}')H_I(\mathbf{r})$ satisfying

$$\int d\mathbf{r}|H_I(\mathbf{r})|^2 < \infty, \tag{1.30}$$

every state $|\phi\rangle$ is the asymptotic state of some state $|\psi\rangle$ in the Hilbert space. That is, in the weak sense, for every $|\phi(t_0)\rangle$ there exist two states $|\psi_\pm(t_0)\rangle$ such that

$$\lim_{t\to\pm\infty}\left[U(t, t_0)|\psi_\pm(t_0)\rangle - U_0(t, t_0)|\phi(t_0)\rangle\right] = 0. \tag{1.31}$$

To prove this, consider the state $|\varphi_\pm(t)\rangle$ that results from applying $U^\dagger(t, t_0)$ from the left on the expression in brackets in equation (1.31), which is simply

$$|\varphi_\pm(t)\rangle = |\psi_\pm(t_0)\rangle - U^\dagger(t, t_0)U_0(t, t_0)|\phi(t_0)\rangle. \tag{1.32}$$

Since $|\psi_{\pm}(t_0)\rangle$ is in the Hilbert space by construction, proving our assertion on asymptotic behaviour requires showing that $U^{\dagger}(t, t_0)U_0(t, t_0)|\phi(t_0)\rangle$ is also in the Hilbert space as $t \to \pm\infty$. That is, if $|\phi(t_0)\rangle$ is in the Hilbert space, the state $|\Phi(t)\rangle = U^{\dagger}(t, t_0)U_0(t, t_0)|\phi(t_0)\rangle$ remains in the Hilbert space in the limit $t \to \pm\infty$. If this is the case, then we can identify $|\psi_{\pm}(t_0)\rangle$ with the limit $\lim_{t\to\pm\infty}|\Phi(t)\rangle$. Differentiating $U^{\dagger}U_0$ with respect to time, we easily prove that

$$\partial_t[U^{\dagger}(t, t_0)U_0(t, t_0)] = \frac{i}{\hbar}U^{\dagger}(t, t_0)H_I U(t, t_0), \tag{1.33}$$

whose formal solution is given by

$$U^{\dagger}(t, t_0)U_0(t, t_0) = 1 + \frac{i}{\hbar}\int_{t_0}^{t} dt' U^{\dagger}(t', t_0)H_I U_0(t', t_0), \tag{1.34}$$

where we have used that $U(t_0, t_0) = U_0(t_0, t_0) = 1$. A sufficient condition for the limits $t \to \pm\infty$ of equation (1.34) to exist is that for every $|\phi(t_0)\rangle$ in the Hilbert space, it holds that

$$\int_{t_0}^{\pm\infty} dt' ||U^{\dagger}(t', t_0)H_I U_0(t', t_0)|\phi(t_0)\rangle|| < \infty. \tag{1.35}$$

Firstly, we use the fact that $U(t, t_0)$ is unitary, so that $||U^{\dagger}(t', t_0)H_I U_0(t', t_0)|\phi(t_0)\rangle|| = ||H_I U_0(t', t_0)|\phi(t_0)\rangle||$. Then, recall that

$$F(t) \equiv ||H_I U_0(t, t_0)|\phi(t_0)\rangle||^2 = \langle\phi(t_0)|U_0^{\dagger}(t, t_0)H_I^2 U_0(t, t_0)|\phi(t_0)\rangle. \tag{1.36}$$

We now particularize to the non-relativistic, translationally invariant case, and insert two complete sets of momentum states in the last term of equation (1.36), and we have

$$F(t) = \int\frac{d\mathbf{k}}{(2\pi)^D}\int\frac{d\mathbf{k}'}{(2\pi)^D}\langle\phi(t_0)|U_0^{\dagger}(t, t_0)|\mathbf{k}'\rangle\langle\mathbf{k}'|H_I^2|\mathbf{k}\rangle\langle\mathbf{k}|U_0(t, t_0)|\phi(t_0)\rangle. \tag{1.37}$$

Now, since momentum states $|\mathbf{k}\rangle$ are eigenstates of H_0 with eigenvalue $\hbar\omega(\mathbf{k}) = \hbar^2\mathbf{k}^2/m$, equation (1.37) becomes

$$F(t) = \int\frac{d\mathbf{k}}{(2\pi)^D}\int\frac{d\mathbf{k}'}{(2\pi)^D}\phi^*(\mathbf{k}', t_0)\phi(\mathbf{k}, t_0)\langle\mathbf{k}'|H_I^2|\mathbf{k}\rangle \\ \exp[i(\omega(\mathbf{k}) - \omega(\mathbf{k}'))(t - t_0)]. \tag{1.38}$$

Above, the Fourier transform of H_I^2, since H_I is local, can be written as

$$\langle\mathbf{k}'|H_I^2|\mathbf{k}\rangle = \int d\mathbf{r}e^{i(\mathbf{k}-\mathbf{k}')\cdot\mathbf{r}}[H_I(\mathbf{r})]^2. \tag{1.39}$$

Since H_I is a two-body interaction, it must satisfy Newton's third law, so $H_I(\mathbf{r}) = H_I(-\mathbf{r})$, and therefore $\langle\mathbf{k}'|H_I^2|\mathbf{k}\rangle$ is real. Hence, equation (1.39) is bounded as

$$\langle\mathbf{k}'|H_I^2|\mathbf{k}\rangle \leqslant \int d\mathbf{r}[H_I(\mathbf{r})]^2 \equiv C, \tag{1.40}$$

which is a finite quantity by hypothesis. Using the bound (1.40) in equation (1.38), we arrive at the simple expression

$$F(t) \leqslant C \left| \int \frac{d\mathbf{k}}{(2\pi)^D} \phi(\mathbf{k},\, t_0) e^{i\omega(\mathbf{k})(t-t_0)} \right|^2. \tag{1.41}$$

Following Taylor [1], we may use a (non-orthonormal) basis of Gaussian functions for $\phi(\mathbf{k},\, t_0)$. We may utilize $\phi(\mathbf{k},\, t_0) = \prod_{i=1}^{D} \phi_i(k_i)$, with $\phi_i(k) \propto \exp(-k^2/b^2)$. It is a straightforward matter to show that the bound in equation (1.41) for large $|t - t_0|$ behaves as $|t - t_0|^{-D/4}$. For integer D, equation (1.35) is fulfilled for $D \geqslant 3$, while for $D \leqslant 2$ this is not the case. Therefore, we have shown that every state $|\phi\rangle$ is the asymptotic state of some other state $|\psi\rangle$ in the Hilbert space for space dimensions equal to or larger than 3. To prove that this is the case in $D = 1$ and $D = 2$, a tighter bound than equation (1.40) would need to be found. Instead of this, later we will see that scattering is well defined in those dimensions by studying the stationary scattering states.

To continue our path towards the S-matrix, we now define the so-called Møller operators $\Omega_{\mp}$. These are given by

$$\Omega_{\mp} = \lim_{t\to\pm\infty} U^{\dagger}(t,\, t_0) U_0(t,\, t_0). \tag{1.42}$$

The motivation for introducing these comes from equation (1.31). Since $|\psi_{\pm}(t_0)\rangle$ are time independent, then

$$|\psi_{\pm}(t_0)\rangle = \Omega_{\mp} |\phi(t_0)\rangle. \tag{1.43}$$

Now, we assume that the in- and out-asymptotes are given by ϕ_{in} and ϕ_{out}, such that the actual wave function at $t = t_0$ is given by $|\psi(t_0)\rangle$. Then, from equation (1.43), we have

$$|\psi(t_0)\rangle = \Omega_{-} |\phi_{\text{out}}\rangle = \Omega_{+} |\phi_{\text{in}}\rangle. \tag{1.44}$$

Since the Møller operators are isometric, we have that

$$|\phi_{\text{in}}\rangle = \Omega_{+}^{\dagger} |\psi(t_0)\rangle, \tag{1.45}$$

$$|\phi_{\text{out}}\rangle = \Omega_{-}^{\dagger} |\psi(t_0)\rangle. \tag{1.46}$$

Therefore, we finally obtain

$$|\phi_{\text{out}}\rangle = \Omega_{-}^{\dagger} |\psi(t_0)\rangle = \Omega_{-}^{\dagger} \Omega_{+} |\phi_{\text{in}}\rangle \equiv S |\phi_{\text{in}}\rangle, \tag{1.47}$$

where we have defined the scattering operator $S = \Omega_{-}^{\dagger}\Omega_{+}$, which takes the in-asymptote into the out-asymptote. The physical interpretation of the scattering operator is very simple. Assume a scattering experiment prepared, for asymptotically long times in the past ($t \to -\infty$), in the in-asymptote $|\xi\rangle$. The probability that in the experiment some out-asymptote $|\phi\rangle$ is measured is given by the squared modulus of the overlap between the actual wave function that evolves from $|\xi\rangle$ and the wave function that the target out-asymptote $|\phi\rangle$ would evolve from at any moment of time t_0. These respective wave functions are obtained from equation (1.45) and (1.46), as

$$|\psi_1\rangle = \Omega_+|\xi\rangle, \tag{1.48}$$

$$|\psi_2\rangle = \Omega_-|\phi\rangle, \tag{1.49}$$

and, thus, the probability is given by

$$P(\xi \to \phi) = |\langle\psi_1|\psi_2\rangle|^2 = |\langle\xi|\Omega_+^\dagger\Omega_-|\phi\rangle|^2 = |\langle\xi|S|\phi\rangle|^2. \tag{1.50}$$

The probability amplitude is therefore given by the matrix element of the scattering operator, the collection of which is called the S-matrix. Generally, however, we shall use the terms 'scattering operator' and 'S-matrix' interchangeably. A fundamental—yet highly non-trivial—result in scattering theory is the unitarity of the scattering operator, that is, $S^\dagger S = SS^\dagger = 1$.

1.2.2 Green's and transition operators

We introduce here two operators that are of central importance to the quantum theory of scattering, called Green's operator and the transition operator. We follow closely the discussion offered in Taylor's book [1]. The relevance of these operators, which will be made apparent shortly, comes from two facts, one for each operator. Firstly, knowledge of the Green's operator, whose matrix elements are Green's functions, determines the spectrum of the Hamiltonian completely. As for the transition operator, the collection of its elements called the T-matrix determines, in conjunction with the non-interacting Green's function, the scattered wave of a scattering state. On the energy shell (on-shell), the T-matrix also specifies the non-forward scattering operator or the S-matrix. Moreover, as we shall see, knowledge of the transition operator implies knowledge of the Green's operator and vice versa, and therefore both descriptions of scattering theory are complete.

We begin by studying the Green's operator, which is a function of the (complex) energy z. Given a Hamiltonian H, the Green's operator, denoted $G(z)$, is defined as

$$G(z) \equiv (z - H)^{-1}. \tag{1.51}$$

Clearly, $G(z)$ does not exist if $z \in \sigma(H)$, with $\sigma(H)$ the spectrum of the operator H. If the Hamiltonian can be split into non-interacting (H_0) and interacting (V) parts as $H = H_0 + V$, then we call $G_0(z) = (z - H_0)^{-1}$ the non-interacting Green's operator. As mentioned above, the matrix elements of the Green's operator are Green's functions. To see this, we begin with the non-interacting Green's function, and assume that $z \notin \sigma(H_0)$. Therefore, we may invert $G_0(z)$, and write the following equation

$$(z - H_0)G_0(z) = 1. \tag{1.52}$$

In the position representation, equation (1.52) takes the form

$$\langle\mathbf{r}|(z - H_0)G_0(z)|\mathbf{r}'\rangle = \langle\mathbf{r}|\mathbf{r}'\rangle = \delta(\mathbf{r} - \mathbf{r}'). \tag{1.53}$$

If, as usual, the non-interacting Hamiltonian H_0 is a local, differential (or difference, on a lattice) operator, we have

$$\langle \mathbf{r}|H_0|\mathbf{r}'\rangle \equiv \delta(\mathbf{r}-\mathbf{r}')\hat{h}_0(\mathbf{r}), \tag{1.54}$$

where the hat on the operator $\hat{h}_0(\mathbf{r})$ is used to remind us that it is a differential (or difference) operator. We insert a resolution of the identity in equation (1.53), obtaining

$$z\langle \mathbf{r}|G_0(z)|\mathbf{r}'\rangle - \int d\mathbf{r}''\langle \mathbf{r}|H_0|\mathbf{r}''\rangle\langle \mathbf{r}''|G_0(z)|\mathbf{r}'\rangle = \delta(\mathbf{r}-\mathbf{r}'), \tag{1.55}$$

and after inserting equation (1.54) into equation (1.55), and defining $\langle \mathbf{r}|G_0(z)|\mathbf{r}'\rangle \equiv G_0(z; \mathbf{r}, \mathbf{r}')$, we obtain

$$(z - \hat{h}_0(\mathbf{r}))G_0(z; \mathbf{r}, \mathbf{r}') = \delta(\mathbf{r}-\mathbf{r}'), \tag{1.56}$$

which is exactly the equation satisfied by the Green's function for the differential (or difference) operator $\hat{h}_0$ for $z \notin \sigma(H_0)$. The same reasoning applies to the Green's function of the full Hamiltonian H, which is holomorphic (in z) for $z \notin \sigma(H)$. Its analytic structure is also quite clear, having simple poles at the discrete energy eigenvalues $\{E_n\}$, and a branch cut for real $z > 0$ in the non-relativistic case.

Let us work with the standard three-dimensional space, in the non-relativistic case. For short-range interactions, H has continuous spectrum, which is the part of the spectrum that is of interest in scattering theory. For simplicity, we analyse the non-interacting Green's operator $G_0(z)$. Call $|E\ell m\rangle$ the non-interacting continuum states with energy $E > 0$, and angular momentum quantum numbers $\ell = 0, 1, \ldots$ and $m = -\ell, -\ell+1, \ldots, \ell$. The non-interacting Green's operator has the matrix elements

$$\langle E'\ell' m'|G_0(z)|E\ell m\rangle = \frac{1}{z-E}\delta(E-E')\delta_{\ell\ell'}\delta_{mm'}. \tag{1.57}$$

Therefore, it can be written as

$$G_0(z) = \int_0^\infty dE \sum_{\ell m} \frac{|E\ell m\rangle\langle E\ell m|}{z-E}. \tag{1.58}$$

We want to prove now that $G_0(z)$ has a branch cut on the positive real axis, i.e., for $z \in (0, \infty)$. To see this, it suffices to show that this is the case for its proper matrix elements. That is, take two vectors $|\chi\rangle$ and $|\psi\rangle$ in the Hilbert space (normalizable states). The matrix elements are given by

$$\langle \chi|G_0(z)|\psi\rangle = \int_0^\infty \frac{dE}{z-E}\sum_{\ell m}\langle \chi|E\ell m\rangle\langle E\ell m|\psi\rangle \equiv \int_0^\infty \frac{dE}{z-E}A_{E\ell m}(\chi, \psi). \tag{1.59}$$

We now define a certain energy $E_0 > 0$, as well as $z_\pm \equiv E_0 \pm i\eta$ ($\eta \to 0^+$). We shall prove that the matrix elements in equation (1.59) for $z = z_\pm$ do not agree with each other. Hence, we calculate their difference as

$$\langle\chi|G_0(z_+)|\psi\rangle - \langle\chi|G_0(z_-)|\psi\rangle = \int_0^\infty dE\left[\frac{1}{E_0 - E + i\eta} - \frac{1}{E_0 - E - i\eta}\right] A_{E\ell m}(\chi, \psi). \quad (1.60)$$

To calculate the difference above, we make use of Plemelj's formula, which reads

$$\frac{1}{E_0 - E \pm i\eta} = \mathcal{P}\frac{1}{E_0 - E} \mp i\pi\delta(E_0 - E). \quad (1.61)$$

Using equation (1.61) in equation (1.60), we see that

$$\langle\chi|G_0(z_+)|\psi\rangle - \langle\chi|G_0(z_-)|\psi\rangle = -2\pi i A_{E_0\ell m}(\chi, \psi) \neq 0, \quad (1.62)$$

which implies that G_0 has a branch cut on the positive real axis, as we wanted to show. The branch cut is of central importance in scattering theory, since choosing to approach the positive real axis from above or below (in the complex plane) yields different boundary conditions to the stationary scattering states.

We are now going to obtain an equation for the Green's function of the full Hamiltonian H, given $G_0(z)$ and the interaction V. For this, we first show that if A and B are two invertible operators, then

$$A^{-1} = B^{-1} + B^{-1}(B - A)A^{-1}. \quad (1.63)$$

To prove the above, we apply A on the right to both sides of equation (1.63). If we now set $A = z - H$ and $B = z - H_0$, equation (1.63) becomes

$$G(z) = G_0(z) + G_0(z)VG(z), \quad (1.64)$$

which is called the **Lippmann–Schwinger equation** for the Green's function. It is also easy to show that

$$G(z) = G_0(z) + G(z)VG_0(z). \quad (1.65)$$

The above equation is completely equivalent to the Lippmann–Schwinger equation (1.64). A last property, which is also important, of a Green's function for a self-adjoint operator H is that

$$G(z^*) = [G(z)]^\dagger. \quad (1.66)$$

The proof is as follows

$$G(z^*) = (z^* - H)^{-1}, \quad (1.67)$$

$$[G(z)]^\dagger = [(z - H)^{-1}]^\dagger = (z^* - H^\dagger)^{-1} = (z^* - H)^{-1}. \quad (1.68)$$

We now move on to introduce the so-called transition operator. This also depends on the (complex) energy z, and is denoted $T(z)$. The set of its matrix elements will be called the T-matrix. The transition operator, given a Hamiltonian $H = H_0 + V$, is defined via the Green's operator and the interaction as

$$T(z) = V + VG(z)V. \tag{1.69}$$

Our goal is now to obtain a Lippmann–Schwinger equation for the transition operator. To do this, we first derive a number of relations. Apply the non-interacting Green's operator $G_0(z)$ on the left of both sides of equation (1.69), obtaining

$$G_0(z)T(z) = G_0(z)V + G_0(z)VG(z)V = [G_0(z) + G_0(z)VG(z)]V = G(z)V, \tag{1.70}$$

where, in the last step, we have used the Lippmann–Schwinger equation for the Green's operator, equation (1.64). Multiplying equation (1.69) on the right with $G_0(z)$, and using equation (1.65) we obtain instead

$$T(z)G_0(z) = VG(z). \tag{1.71}$$

We now replace, in equation (1.64), $VG(z)$ with $T(z)G_0(z)$, equation (1.71). Then we obtain an inverted relation to obtain $G(z)$ given the transition operator $T(z)$ and the non-interacting Green's function $G_0(z)$, as

$$G(z) = G_0(z) + G_0(z)T(z)G_0(z). \tag{1.72}$$

The relation above implies that if we know $T(z)$, we know $G(z)$, and therefore knowledge of $T(z)$ is equivalent to knowledge of $G(z)$. We now use relation (1.71) in the definition of the transition operator, equation (1.69), obtaining the Lippmann–Schwinger equation for the transition operator,

$$\boxed{T(z) = V + VG_0(z)T(z).} \tag{1.73}$$

The Lippmann–Schwinger equation that we have just obtained will be the central equation to be tackled in this book and, in fact, the only boxed equation the reader will ever encounter within these pages. We will use it extremely often, and the reader is encouraged to learn it by heart.

We are now in position to relate the Green's and transition operators with the quantum theory of scattering. We consider the quantum-mechanical collision of a particle with a target, described by a short-range potential. We shall denote the state of the system at time $t = 0$ $|\psi\rangle$. If the in-asymptote is called $|\phi\rangle$, then $|\psi\rangle$ is obtained from it by applying the Møller operator Ω_+, that is,

$$|\psi\rangle = \Omega_+|\phi\rangle \equiv |\phi+\rangle. \tag{1.74}$$

Above, the '+' sign after ϕ is used to specify that $|\phi+\rangle$ is the result of the application of the Møller operator Ω_+ on the wave function $|\phi\rangle$. This is convenient notation since, if we were to denote $|\phi\rangle$ the *out*- asymptote, then we would have

$$|\psi\rangle = \Omega_-|\phi\rangle \equiv |\phi-\rangle, \tag{1.75}$$

which specifies that $|\phi-\rangle$ is the result of the application of the Møller operator Ω_- on the wave function $|\phi\rangle$. Recall that

$$|\phi-\rangle = |\phi\rangle + \frac{i}{\hbar}\int_0^{\infty} dt' U^{\dagger}(t')VU_0(t')|\phi\rangle. \tag{1.76}$$

We now use a well known result, that is, that if $\exists \int_0^\infty dt' |f(t')\rangle$ (i.e., if it is in the Hilbert space), we can introduce a damping factor and obtain

$$\int_0^\infty dt' |f(t')\rangle = \lim_{\eta \to 0^+} \int_0^\infty dt' e^{-\eta t'} |f(t')\rangle. \tag{1.77}$$

Obviating the limit $\eta \to 0^+$, which will be assumed implicitly, we can work out equation (1.76) by inserting a complete set of momentum eigenstates, as well as expression (1.77), obtaining

$$\begin{aligned} |\phi-\rangle &= |\phi\rangle + \frac{i}{\hbar} \int \frac{d\mathbf{k}}{(2\pi)^D} \int_0^\infty dt' [e^{-\eta t'} U^\dagger(t') V U_0(t')] |\mathbf{k}\rangle\langle\mathbf{k}|\phi\rangle \\ &= |\phi\rangle + \frac{i}{\hbar} \int \frac{d\mathbf{k}}{(2\pi)^D} \int_0^\infty dt' [e^{-\eta t'} e^{iHt'/\hbar} V e^{-i\epsilon(\mathbf{k})t'/\hbar}] \phi(\mathbf{k}) |\mathbf{k}\rangle \\ &= |\phi\rangle + \frac{i}{\hbar} \int \frac{d\mathbf{k}}{(2\pi)^D} \int_0^\infty dt' \exp\left[-\frac{i}{\hbar}\left(\epsilon(\mathbf{k}) - H + i\hbar\eta \right) t' \right] \phi(\mathbf{k}) V |\mathbf{k}\rangle. \end{aligned} \tag{1.78}$$

We use the following trivial integral $\int_0^\infty dt' \exp(-\eta t' - i\alpha t') = 1/(\eta + i\alpha)$ $(\alpha \in \mathbb{R})$, and finally obtain (the calculation is analogous for $|\phi + \rangle$)

$$|\phi\pm\rangle = |\phi\rangle + \int \frac{d\mathbf{k}}{(2\pi)^D} \frac{1}{\epsilon(\mathbf{k}) - H \pm i\eta} \phi(\mathbf{k}) V |k\rangle. \tag{1.79}$$

If we now take the plane-wave limit that will define stationary scattering states, that is, $|\phi\rangle = |\mathbf{q}\rangle$ for some momentum $\mathbf{q}$, and identify the Green's function in the integrand of equation (1.79), we obtain

$$|\mathbf{q}\pm\rangle = |\mathbf{q}\rangle + G(\epsilon(\mathbf{q}) \pm i\eta) V |\mathbf{q}\rangle = |\mathbf{q}\rangle + T(\epsilon(\mathbf{q}) \pm i\eta) G_0(\epsilon(\mathbf{q}) \pm i\eta) |\mathbf{q}\rangle. \tag{1.80}$$

To obtain the above, we have used $\phi(\mathbf{k}) = (2\pi)^D \delta(\mathbf{k} - \mathbf{q})$, and then used equation (1.71) to write it in terms of the transition operator. Since it is easy to see that $|\mathbf{q}\pm\rangle$ are stationary continuum states of the Schrödinger equation, that is, stationary scattering states, we call $|\mathbf{q}\rangle$ the incident wave, and $|\psi_{sc}\rangle$, defined as

$$|\psi_{sc}\rangle \equiv T(\epsilon(\mathbf{q} \pm i\eta)) G_0(\epsilon(\mathbf{q} \pm i\eta)) |\mathbf{q}\rangle, \tag{1.81}$$

the scattered wave.

We continue our quest to give physical value to the transition operator by studying the scattering operator via its matrix elements (the S-matrix). Our starting point is its definition via the Møller operators $\Omega_\pm$, which we saw (see equation (1.47)) is given by

$$S = \Omega_-^\dagger \Omega_+. \tag{1.82}$$

Taking again two proper vectors in the Hilbert space, $|\chi\rangle$ and $|\phi\rangle$, the S-matrix element reads

$$\langle\chi|S|\phi\rangle = \langle\chi|\Omega_-^\dagger\Omega_+|\phi\rangle = \lim_{t,-t' \to\infty} \langle\chi|e^{iH_0t/\hbar}e^{-iHt/\hbar}e^{iHt'/\hbar}e^{-iH_0t'/\hbar}|\phi\rangle. \tag{1.83}$$

Since the limit above exists, we can simultaneously take both limits by setting $t' = -t$ and taking $t \to \infty$. Equation (1.83) simplifies to

$$\langle\chi|S|\phi\rangle = \lim_{t\to\infty}\langle\chi|e^{iH_0t/\hbar}e^{-2iHt/\hbar}e^{iH_0t/\hbar}|\phi\rangle \equiv \langle\chi|\hat{O}(t)|\phi\rangle. \tag{1.84}$$

We differentiate the operator $\hat{O}$ above with respect to time, which gives

$$\frac{d\hat{O}}{dt}(t) = \frac{iH_0}{\hbar}\hat{O}(t) - \frac{2i}{\hbar}e^{iH_0t/\hbar}He^{-2iHt/\hbar}e^{iH_0t/\hbar} + \frac{i}{\hbar}\hat{O}(t)H_0. \tag{1.85}$$

Using that $H_0 = (H - V)$, equation (1.85) reduces to

$$\frac{d\hat{O}}{dt}(t) = -\frac{i}{\hbar}[e^{iH_0t/\hbar}Ve^{-2iHt/\hbar}e^{iH_0t/\hbar} + e^{iH_0t/\hbar}e^{-2iHt/\hbar}Ve^{iH_0t/\hbar}]. \tag{1.86}$$

Initially (at $t = 0$), the operator $\hat{O}$ is the identity, that is, $\hat{O}(0) = 1$. Therefore, we can solve equation (1.86) and write formally

$$\hat{O}(t) = 1 - \frac{i}{\hbar}\int_0^t dt'[e^{iH_0t'/\hbar}Ve^{-2iHt'/\hbar}e^{iH_0t'/\hbar} + e^{iH_0t'/\hbar}e^{-2iHt'/\hbar}Ve^{iH_0t'/\hbar}]. \tag{1.87}$$

Inserting the above relation into the S-matrix elements (1.84), and inserting the exponential damping factor $\exp(-\eta t)$, we have

$$\begin{aligned}\langle\chi|S|\phi\rangle &= \langle\chi|\phi\rangle \\ &- \frac{i}{\hbar}\int_0^\infty dt\langle\chi|[e^{iH_0t/\hbar}Ve^{-2iHt/\hbar}e^{iH_0t/\hbar} + e^{iH_0t/\hbar}e^{-2iHt/\hbar}Ve^{iH_0t/\hbar}]|\phi\rangle e^{-\eta t},\end{aligned} \tag{1.88}$$

where the limit $\eta \to 0^+$ is implicit, as usual. Thanks to the damping factor in the time integral above, we can go on to use improper states, that is, plane waves. Equation (1.88) for these continuum states takes the form

$$\begin{aligned}\langle\mathbf{k}'|S|\mathbf{k}\rangle &= (2\pi)^D\delta(\mathbf{k} - \mathbf{k}') \\ &- \frac{i}{\hbar}\int_0^\infty dte^{-i(\epsilon(\mathbf{k}')-\epsilon(\mathbf{k}))t/\hbar}\langle\mathbf{k}'|[Ve^{-2iHt/\hbar} + e^{-2iHt/\hbar}V]|\mathbf{k}\rangle e^{-\eta t}.\end{aligned} \tag{1.89}$$

Performing the integral over time, and identifying the corresponding Green's functions, equation (1.89) becomes

$$\langle\mathbf{k}'|S - 1|\mathbf{k}\rangle = \frac{1}{2\hbar}\langle\mathbf{k}'|\left[VG\left(\frac{\epsilon(\mathbf{k}') - \epsilon(\mathbf{k}')}{2} + i\eta\right) + G\left(\frac{\epsilon(\mathbf{k}) - \epsilon(\mathbf{k}')}{2} + i\eta\right)V\right]|\mathbf{k}\rangle. \tag{1.90}$$

Using relations (1.70) and (1.71) to write equation (1.90) above in terms of the T-matrix, and then applying Plemelj's formula, equation (1.61), we finally obtain

$$\langle\mathbf{k}'|(S - 1)|\mathbf{k}\rangle = -\frac{2\pi i}{\hbar}\delta(\epsilon(\mathbf{k}) - \epsilon(\mathbf{k}'))\langle\mathbf{k}'|T(\epsilon(\mathbf{k} + i\eta)|\mathbf{k}\rangle. \tag{1.91}$$

The above relation is also central in scattering theory. It tells us that the (non-trivial part of) the S-matrix is given by the T-matrix on the energy shell. That is, the on-shell T-matrix, which is enforced by the fact that $\epsilon(\mathbf{k}) = \epsilon(\mathbf{k}')$. In non-relativistic, Galilean scattering, this simply implies that $|\mathbf{k}| = |\mathbf{k}'|$.

1.2.3 Partial wave scattering

Collision theory as we have studied so far is completely general. That is, it is independent of symmetries or lack thereof. However, symmetry plays a very important role in the quantum theory of scattering. The most important and general symmetry that we may have in non-relativistic, Galilean scattering, is spherical symmetry when two-body interactions are central. In this case, the Hamiltonian ($H = H_0 + V$) commutes with L^2 and L_z, and we use the basis of non-interacting eigenstates, i.e., eigenstates of H_0, that we label with their energy $E = \hbar^2k^2/2\mu$, with μ the reduced mass, and their angular momentum quantum numbers ℓ and m. In this basis, the S-matrix takes the form

$$\langle E'\ell'm'|S|E\ell m\rangle = \delta(E' - E)\delta_{\ell\ell'}\delta_{mm'}s_\ell(k). \tag{1.92}$$

Above, $s_\ell(k)$ is a complex number to be determined, and is known as the partial wave S-matrix. The specific form of $s_\ell(k)$ can be found by recalling that the scattering operator is unitary, $S^\dagger S = 1$. In the basis of non-interacting states with good angular momentum quantum numbers, unitarity is expressed as

$$\begin{aligned}\langle E'\ell'm'|S^\dagger S|E\ell m\rangle &= \int dE'' \sum_{\ell''m''}\langle E'\ell'm'|S^\dagger|E''\ell''m''\rangle\langle E''\ell''m''|S|E\ell m\rangle \\ &= \delta(E' - E)\delta_{\ell\ell'}\delta_{mm'}.\end{aligned} \tag{1.93}$$

Using equation (1.92) in equation (1.93) above, we immediately obtain the following result

$$|s_\ell(k)|^2 = 1 \Rightarrow s_\ell(k) \equiv e^{2i\delta_\ell(k)}, \tag{1.94}$$

where $\delta_\ell(k)$ is a real number that we call the partial wave scattering phase shift. After introducing the scattering phase shifts, our goal switches to finding a relation between the on-shell T-matrix, which we can calculate using the Lippmann–Schwinger equation (1.73), and the scattering phase shifts. Firstly, we use equation (1.92) to write

$$(S - 1)|E\ell m\rangle = (s_\ell(k) - 1)|E\ell m\rangle. \tag{1.95}$$

Writing the on-shell T-matrix as $t(\mathbf{k}, \mathbf{k}')$ for notational simplicity, equation (1.95) together with equation (1.91) allows us to write

$$-\frac{2\pi i}{\hbar}\delta(\epsilon(\mathbf{k}) - \epsilon(\mathbf{k}'))t(\mathbf{k}, \mathbf{k}') = \int dE\sum_{\ell m}(s_\ell(k) - 1)\langle\mathbf{k}'|E\ell m\rangle\langle E\ell m|\mathbf{k}\rangle. \tag{1.96}$$

All that remains now is the calculation of the momentum representation of the states $|E\ell m\rangle$. In the position representation, these are well known, and given by

$$\langle \mathbf{r}|E\ell m\rangle = i^{\ell}\left(\frac{2\mu k}{\pi}\right)^{1/2} j_{\ell}(kr)Y_{\ell m}(\hat{r}), \tag{1.97}$$

where j_ℓ is the spherical Bessel function of order ℓ, and $Y_{\ell m}$ is a spherical harmonic. We will make use of the expansion of a plane wave in terms of spherical Bessel functions and spherical harmonics, namely

$$\langle \mathbf{r}|\mathbf{k}\rangle = e^{i\mathbf{k}\cdot\mathbf{r}} = 4\pi\sum_{\ell m} i^{\ell} j_{\ell}(kr)Y^{*}_{\ell m}(\hat{k})Y_{\ell m}(\hat{r}), \tag{1.98}$$

as well as the k-space orthogonality relation between spherical Bessel functions,

$$\int_0^{\infty} dr r^2 j_{\ell}(kr) j_{\ell}(k'r) = \frac{\pi}{2k^2}\delta(k - k'), \tag{1.99}$$

and the orthonormality of spherical harmonics. All these, inserted into

$$\langle \mathbf{k}|E\ell m\rangle = \int d\mathbf{r}\langle \mathbf{k}|\mathbf{r}\rangle\langle \mathbf{r}|E\ell m\rangle, \tag{1.100}$$

yield the following expression for the momentum representation of $|E\ell m\rangle$

$$\langle \mathbf{k}'|E\ell m\rangle = \frac{4\pi}{\hbar}\left(\frac{2\mu k}{\pi}\right)^{1/2} Y^{*}_{\ell m}(\hat{k})\frac{\pi}{2k^2}\delta(k - k'). \tag{1.101}$$

We finally write $\delta(k - k')$ as a Dirac delta function for energies, using $\delta(k - k') = (\hbar^2 k/\mu)\delta(E - \epsilon(\mathbf{k}'))$, and equation (1.101) finally becomes

$$\langle \mathbf{k}|E\ell m\rangle = \hbar\frac{(2\pi)^{3/2}}{(\mu k)^{1/2}} Y_{\ell m}(\hat{k})\delta(E - \epsilon(\mathbf{k})). \tag{1.102}$$

We insert equation (1.102) into equation (1.96), and we easily obtain, for the on-shell T-matrix,

$$-2\pi i t(\mathbf{k}, \mathbf{k}') = \frac{(2\pi)^3\hbar^2}{\mu k}\sum_{\ell m} Y^{*}_{\ell m}(\hat{k}')Y_{\ell m}(\hat{k})(s_{\ell}(\epsilon(\mathbf{k})) - 1). \tag{1.103}$$

We can now get rid of the sum over the z-component of the orbital angular momentum by using the addition theorem for the spherical harmonics,

$$\sum_{m=-\ell}^{\ell} Y^{*}_{\ell m}(\hat{k}')Y_{\ell m}(\hat{k}) = \frac{2\ell + 1}{4\pi}P_{\ell}(\cos\theta), \tag{1.104}$$

where θ is the angle between $\mathbf{k}$ and $\mathbf{k}'$ and $P_\ell(x)$ a Legendre polynomial, in equation (1.103). We finally end up with the partial wave expansion for the on-shell T-matrix, which reads

$$t(\mathbf{k}, \mathbf{k}') = \sum_{\ell=0}^{\infty}(2\ell + 1)t_{\ell}(\epsilon(\mathbf{k}))P_{\ell}(\cos\theta), \tag{1.105}$$

$$t_\ell(\epsilon(\mathbf{k})) = \frac{i\pi\hbar^2}{\mu k}[s_\ell(\epsilon(\mathbf{k})) - 1]. \tag{1.106}$$

It is also convenient to write the partial wave on-shell T-matrix in terms of the phase shifts δ_ℓ, defined in equation (1.94). It is straightforward to check that equation (1.106) takes the form

$$t_\ell(\epsilon(\mathbf{k})) = -\frac{2\pi\hbar^2}{\mu k}e^{i\delta_\ell(k)}\sin\delta_\ell(k). \tag{1.107}$$

Writing the above in terms of sine and cosine functions only, and multiplying and dividing by the complex conjugate, we obtain

$$t_\ell(\epsilon(\mathbf{k})) = -\frac{2\pi\hbar^2}{\mu k}\frac{1}{\cot\delta_\ell(k) - i}, \tag{1.108}$$

which is the usual and most convenient form of writing the on-shell T-matrix in terms of the scattering phase shifts. In particular, this expression is very useful for low-energy collision theory. It is also the starting point of the so-called effective range expansions, which we build on next.

1.2.4 Effective range expansions

We study an extremely important tool for the theory of low-energy interactions that we will develop in this book. This is called the effective range expansion (or effective range expansions if non-zero angular momenta are considered). Essentially, it consists of a power series in the collision energy for the partial wave scattering phase shifts or, more concretely, for the quantity $k\cot\delta_\ell(k)$. Sometimes, instead of $k\cot\delta_\ell(k)$ we shall refer to *the real part of the inverse of the on-shell T-matrix*, since, from equation (1.108), we see that

$$-k\cot\delta_\ell(k) = \frac{2\pi\hbar^2}{\mu}\,\mathrm{Re}\left(\frac{1}{t_\ell(\mathbf{k})}\right). \tag{1.109}$$

This wording might (and does) sound cumbersome, but it is rather standard in the literature.

Effective range expansions have a rigorous basis in the theory of Jost functions [1]. However, we shall follow a more practical route that makes use of Wronskians, although we will not refer to them explicitly. We begin with the stationary Schrödinger equation for a non-relativistic two-body system in the relative coordinate, with a central interaction $V(r)$ between the particles, at positive energies, i.e., for scattering states. These can be written as

$$\psi(\mathbf{r}) = ru(r)Y_{\ell m}(\hat{r}). \tag{1.110}$$

Inserting this into the Schrödinger equation at energy $E = \hbar^2k^2/2\mu$, and defining $U_\ell(r) = 2\mu V(r)/\hbar^2 + \ell(\ell+1)/r^2$, the reduced radial wave function $u(r)$ satisfies

$$u''(r) + [k^2 - U_\ell(r)]u(r) = 0. \tag{1.111}$$

We also introduce asymptotic states v that are inadmissible solutions (because of their behaviour at the origin) to equation (1.111) with no interaction, that is, with $U(r) = 0$. For these, we write

$$v''(r) + \left(k^2 - \frac{\ell(\ell+1)}{r^2}\right)v(r) = 0. \tag{1.112}$$

Defining $x = kr$ and $v(r) = x^{1/2}y(x)$, equation (1.112) reduces to the following differential equation

$$x^2y''(x) + xy'(x) + \left[x^2 - \left(\ell + \frac{1}{2}\right)^2\right]y(x) = 0, \tag{1.113}$$

which corresponds to Bessel's differential equation. Its two independent solutions in this case are $y_r(x) = J_{\ell+1/2}(x)$ and $y_i(x) = Y_{\ell+1/2}(x)$, where the subscript r (i) refers to the regular (irregular) solution at the origin. The Bessel functions J and Y of half-integer order are related to spherical Bessel functions j_ℓ and y_ℓ as follows

$$j_\ell(x) = \sqrt{\frac{\pi}{2x}}J_{\ell+1/2}(x), \tag{1.114}$$

$$y_\ell(x) = \sqrt{\frac{\pi}{2x}}Y_{\ell+1/2}(x). \tag{1.115}$$

Hence, the asymptotic states take the form

$$v(r) \propto kr\left[j_\ell(kr) + B_\ell(k)y_\ell(kr)\right], \tag{1.116}$$

where $B_\ell(k)$ is a constant. Before we continue, it is important to look at the asymptotic behaviour ($kr \to \infty$) of the asymptotic states $v(r)$. These behave as

$$v(r) \propto \sin\left(kr - \frac{\ell\pi}{2}\right) - B_\ell(k)\cos\left(kr - \frac{\ell\pi}{2}\right), \quad kr \to \infty. \tag{1.117}$$

We analyse the s-wave ($\ell = 0$) case, which is the dominant channel for spinless bosons at low energies. The phase shifts can be inferred from the asymptotic relation

$$v(r) \propto \sin\left(kr - \frac{\ell\pi}{2} + \delta_\ell(k)\right), \quad kr \to \infty \tag{1.118}$$

Comparing equations (1.118) and (1.117), we see that

$$B_\ell(k) = -\cot\delta_\ell(k). \tag{1.119}$$

We will consider the asymptotic states having the normalization condition

$$v(r) \to \frac{\sin\left(kr - \frac{\ell\pi}{2} + \delta_\ell(k)\right)}{\sin\left(-\frac{\ell\pi}{2} + \delta_\ell(k)\right)}, \quad kr \to \infty, \tag{1.120}$$

which means that, at all values of r (see equation (1.116)), and for $\ell = 0$,

$$v(r) = kr\left[j_0(kr)\cot\delta_0(k) - y_0(kr)\right]. \tag{1.121}$$

The actual scattering states $u(r)$, satisfying equation (1.111), behave asymptotically like $v(r)$. Therefore, we can impose on them the same normalization. Consider two different collisional energies, $E_i = \hbar^2 k_i^2/2\mu$ ($i = 1, 2$), and call $u_i(r)$ the respective partial wave scattering states with the same angular momentum quantum number ℓ. Then, multiply equation (1.111) for u_1 (u_2) by u_2 (u_1), and subtract the resulting equations from each other. We obtain

$$(u_1 u_2'' - u_2 u_1'') + (k_2^2 - k_1^2)u_1 u_2 = 0. \tag{1.122}$$

Integrating equation (1.122) from 0 to a fixed $R > 0$, and integrating by parts, we get

$$(u_1 u_2')_0^R - (u_2 u_1')_0^R = \int_0^R dr(k_1^2 - k_2^2)u_1 u_2. \tag{1.123}$$

We use $u_i(0) = 0$, and finally obtain

$$u_1(R)u_2'(R) - u_2(R)u_1'(R) = (k_1^2 - k_2^2)\int_0^R dr u_1(r)u_2(r). \tag{1.124}$$

We do the very same for the asymptotic states v_i, with the same collisional energies and angular momenta as for the full scattering states u_i, that is

$$(v_1 v_2')_0^R - (v_2 v_1')_0^R = \int_0^R dr(k_1^2 - k_2^2)v_1 v_2. \tag{1.125}$$

In this case, however, it is not true that $v_i(0) = 0$. We subtract equation (1.125) from equation (1.124) and take the limit $R \to \infty$, in which case $u_i(R) - v_i(R) \to 0$, by construction. We can then write

$$\lim_{r\to 0}[v_1(r)v_2'(r) - v_2(r)v_1'(r)] = (k_1^2 - k_2^2)\int_0^\infty [u_1(r)u_2(r) - v_1(r)v_2(r)]. \tag{1.126}$$

Setting $k_1 = 0$, $k_2 = k$ and defining the s-wave scattering length a as

$$\lim_{k_1\to 0} k\cot\delta_0(k_1) = -\frac{1}{a}, \tag{1.127}$$

we obtain

$$-k\cot\delta_0(k) = \frac{1}{a} + k^2\int_0^\infty dr[u_1(r)u_2(r) - v_1(r)v_2(r)]. \tag{1.128}$$

To leading order in k, the above expression simplifies to

$$-k \cot \delta_0(k) = \frac{1}{a} + k^2 \int_0^\infty dr[u_1(r)^2 - v_1(r)^2] + O(k^4), \tag{1.129}$$

where the subscript 1 refers to $k_1 = 0$. We identify the effective range as the following quantity

$$r = -2 \int_0^\infty dr[u_1(r)^2 - v_1(r)^2], \tag{1.130}$$

so that the effective range expansion takes the form

$$-k \cot \delta_0(k) = \frac{1}{a} - \frac{1}{2} r k^2 + O(k^4). \tag{1.131}$$

For $\ell = 1$, however, we cannot use the above formalism directly, since the equivalent expressions diverge (as $r \to 0$). In this case, the expressions become slightly more cumbersome, and the effective range expansion reads, instead

$$-k^3 \cot \delta_1(k) = \frac{1}{a_p} - \frac{1}{2} r_p k^2 + O(k^4), \tag{1.132}$$

where a_p is the p-wave scattering volume, while r_p is the p-wave range parameter, with units of inverse length.

Problems

1. (Inspired by Taylor [1].) Consider a quantum state $|\psi\rangle$ of a particle, and define by $|\psi_\mathbf{a}\rangle$ the same state translated in space by the vector $\mathbf{a}$, i.e., the corresponding wave function is $\langle \mathbf{x}|\psi_\mathbf{k}\rangle = \psi(\mathbf{x} - \mathbf{a})$.
   (a) Show that $|\psi_\mathbf{k}\rangle = \exp(-i\mathbf{p} \cdot \mathbf{a}/\hbar)|\psi\rangle$. (Hint: Fourier transform.)
   (b) Prove that for any other fixed state $|\phi\rangle$, we have $\langle \phi|\psi_\mathbf{a}\rangle \to 0$ for $\mathbf{a} \to \infty$, even though $\psi_\mathbf{a}$ is non-zero. Give a physical interpretation of this result. (Hint: Start with the case of a localized wave function, i.e., one with support in a finite region of space. Then consider the case of a plane wave as an example of a completely delocalized state.)
2. (Inspired by Taylor [1].) Assume that we have an interaction of the form $H_I = |\phi\rangle\langle\phi|$ for some fixed state in the Hilbert space of a single particle. Prove that equation (1.31) is satisfied. This type of potential is called 'separable', since you may easily show that $\langle \mathbf{x}|H_I|\mathbf{y}\rangle = \phi(\mathbf{x})\phi^*(\mathbf{y})$.
3. Consider the Lippmann–Schwinger operator equation

$$T(z) = V + VG_0(z)T(z), \tag{1.133}$$

   and assume that the potential V is constant in *momentum space*, i.e., $\langle \mathbf{p}|V|\mathbf{p}'\rangle = g$.

(a) Show that in momentum space we have

$$\langle \mathbf{p}|T(z)|\mathbf{p}'\rangle = g + g\int \mathrm{d}^3\mathbf{q}\frac{1}{z - \dfrac{\hbar^2\mathbf{q}^2}{2m}}\langle \mathbf{q}|T(z)|\mathbf{p}'\rangle. \tag{1.134}$$

(b) Assume that the matrix element $\langle \mathbf{p}|T(z)|\mathbf{p}'\rangle$ is constant in momentum space (this is a fair assumption for low-energy scattering in cold atoms for instance). Put a cut-off on the momentum integral at $|\mathbf{q}| = \Lambda$ and show that

$$\langle \mathbf{p}|T(z)|\mathbf{p}\rangle = \frac{g}{1 - gI_0(\mathbf{p}, \Lambda)}, \tag{1.135}$$

where the

$$I_0(\mathbf{p}, \Lambda) = -\frac{m}{\pi^2\hbar^2}\left(\Lambda + \frac{|\mathbf{p}|}{2}\log\left(\frac{\Lambda - |\mathbf{p}|}{\Lambda + |\mathbf{p}|}\right) + i\pi\frac{|\mathbf{p}|}{2}\right). \tag{1.136}$$

(Hint: You can use Plemelj's formula and partial fractions to do the integral.)

4. Consider the T-matrix obtained in the previous problem and write for simplicity $z = E$ and $T(E) = \langle \mathbf{p}|T(z)|\mathbf{p}\rangle$ with the understanding that we are on-shell, i.e., $E = \hbar^2|\mathbf{p}|^2/2\mu$. Consider the limit of low-energy scattering to leading order, keeping only the scattering length a in the s-wave. Taking the limit of large (but finite) Λ in the formula of the previous problem, dropping the logarithmic term, and show that you obtain a relation between a and g of the form

$$g = \frac{2\pi\hbar^2}{\mu}\frac{a}{1 - \dfrac{2\Lambda a}{\pi}}. \tag{1.137}$$

5. The T-matrix is often not exactly solvable and a perturbative approach can then be explored, and a famous such expansion is the Born series. In order to derive it, consider a potential of the form λV. Here we assume that there is a real parameter λ that controls the strength of the potential, i.e., we may expand in this parameter, and then set $\lambda = 1$ in the end to obtain the series in terms of V.

(a) Rewrite the Lippmann–Schwinger equation into the form

$$T(z) = (1 - \lambda VG_0(z))^{-1}\lambda V, \tag{1.138}$$

under the assumption that the inverse on the right-hand side exists.

(b) Expand the inverse and show that

$$T(z) = \lambda V + \lambda VG_0(z)\lambda V + \lambda VG_0(z)\lambda VG_0(z)\lambda V + \cdots \tag{1.139}$$

(c) Take the matrix element in momentum space $\langle \mathbf{p}|T(z)|\mathbf{p}'\rangle$ and show that to first order we have

$$\langle \mathbf{p}|T^1(z)|\mathbf{p}'\rangle = \langle \mathbf{p}|V|\mathbf{p}'\rangle = \int \mathrm{d}\mathbf{x} e^{-i\mathbf{p}\cdot\mathbf{x}/\hbar} V(\mathbf{x}) e^{i\mathbf{p}'\cdot\mathbf{x}/\hbar}, \tag{1.140}$$

concluding that the first order Born approximation for the T-matrix (and hence scattering amplitude as well) is the Fourier transform of the potential evaluated at the *momentum transfer* $\mathbf{q} = \mathbf{p} - \mathbf{p}'$.

Reference

[1] Taylor J R 2006 *Scattering Theory: The Quantum Theory of Nonrelativistic Collisions* (New York: Dover)

Chapter 2

Modern scattering theory

Almost any of the great textbooks available on scattering theory continues to explain how a collision experiment measures scattering cross sections for different events, and relates the relevant S-matrix elements, or the scattering amplitudes (on-shell T-matrix), to those cross sections. We highly recommend Taylor's book in this respect [1]. In this volume and the volume that will follow, however, our motivation does not lie in (although discussions can be extended to) the theoretical description of collision experiments. We are mostly interested in applications of scattering theory in few-body physics and effective interactions in few- and many-body physics. Therefore, we consider a method that is having huge success in the theoretical and numerical description of collisions in nuclear physics and quantum chromodynamics, as well as ultracold atomic and condensed matter physics recently. The method allows for an estimation of low-energy scattering parameters and binding energies from exact diagonalisation, in the low-energy sector, of the system's Hamiltonian, and is originally due to Lüscher [2]. Lacking another term for it, we have decided to simply label this and related methods 'modern scattering theory'.

2.1 Elastic scattering from finite size effects

There are many instances in which solving for scattering states or bound states in infinite volume, using the standard techniques of scattering theory, becomes exceedingly complicated. As an illustrative example, we can think of elastic collisions between two composite objects with non-trivial internal structure. If the two objects are made of n and m 'elementary' particles, we need to solve for scattering states of $n + m$ particles. It is often simpler to obtain numerically accurate results of the low-lying spectrum of the corresponding Hamiltonian in a container with finite volume. In this section, we explore the relations between the low-energy eigenvalues of a Hamiltonian and the low-energy scattering properties of the quantum system of interest.

doi:10.1088/978-0-7503-3087-9ch2

We consider two particles of equal mass m in D dimensions whose dynamics are governed by a Hamiltonian $H = H_0 + V$, where H_0 is the non-interacting Hamiltonian and V is an interaction potential. The two-body T-matrix, after separation of centre-of-mass and relative coordinates, is denoted by $\hat{T}_K$, where a parametric dependence on the total momentum or quasi-momentum $K = k_1 + k_2$ is required when the system is not Galilean relativistic, e.g., for two particles on a crystal. We shall place the two-body system in a box with periodic boundary conditions and all sides of equal length L, unless otherwise stated, so that its volume is L^D.

2.1.1 Weakly-coupled two-body scattering I: *s*-wave collisions

We begin by deriving the weak-coupling results for two spinless bosons in three dimensions, as was done in the seminal work of Lüscher [2]. We will, however, use a different method. In particular, we will show that the energy E_0 of the lowest unbound state of two non-relativistic (with relative kinetic energy dispersion $\epsilon(\mathbf{k}) = \hbar^2 k^2/m$) bosons in three dimensions takes the form

$$E_0 = \frac{4\pi\hbar^2 a}{mL^3}\left\{1 + c_1\frac{a}{L} + c_2\frac{a^2}{L^2}\right\} + O(L^{-6}), \tag{2.1}$$

where a is the s-wave scattering length, expression (2.1) is valid for $|a|/L \ll 1$, and c_1 and c_2 are numerical constants with values $c_1 = 2.837\,297\ldots$ and $c_2 = 6.375\ldots$.

We begin by writing the Schrödinger equation for an eigenstate $|\psi\rangle$ with energy E as

$$|\psi\rangle = G_0(L, E)V|\psi\rangle. \tag{2.2}$$

Above, $G_0(L, E) = (E - H_0)^{-1}$ is the non-interacting Green's function on the cube with periodic boundary conditions. Formally, it appears identical to the non-interacting Green's function $G_0(z)$ in free space. However, their difference, lying in the domain of the Hamiltonian, is very important, as we shall see. In this section, and the sections where finite size and infinite size quantities coexist, we need extra notation to avoid confusion. We will use lowercase symbols to denote periodic, normalized momentum states $|\mathbf{k}\rangle$, i.e.,

$$\langle \mathbf{k}'|\mathbf{k}\rangle = \delta_{\mathbf{k},\mathbf{k}'}, \tag{2.3}$$

where $\mathbf{k} = (n_x, n_y, n_z)2\pi/L$, with n_i integers. We will use capital letters to denote momentum states in free space $|\mathbf{Q}\rangle$, i.e.,

$$\langle \mathbf{Q}'|\mathbf{Q}\rangle = (2\pi)^3\delta(\mathbf{Q} - \mathbf{Q}'). \tag{2.4}$$

As such, we have $1 = \sum_{\mathbf{k}}|\mathbf{k}\rangle\langle\mathbf{k}|$ and $1 = (2\pi)^{-3}\int d^3Q|\mathbf{Q}\rangle\langle\mathbf{Q}|$. Also, the identities in both problems are different, and are those corresponding to their particular Hilbert spaces.

If we are interested in perturbative calculations, the Bethe–Goldstone equation is very useful, and we derive it in the following. To do this, we write the following equation for some vector $|\bar{\psi}\rangle$ in the Hilbert space (for the cube with periodic boundary conditions),

$$|\bar{\psi}\rangle = |\psi_0\rangle + [G_0(L, E)V - |\psi_0\rangle\langle\Gamma|]|\bar{\psi}\rangle, \tag{2.5}$$

where $|\psi_0\rangle$ and $|\Gamma\rangle$ are arbitrary vectors in the Hilbert space. For those values of the energy for which $\langle\Gamma|\bar{\psi}\rangle = 1$, it holds that $|\bar{\psi}\rangle = |\psi\rangle$, i.e., it satisfies equation (2.2). To make equation (2.5) useful, we define $|\psi_0\rangle = |\mathbf{k}\rangle$ (a momentum state), and $\langle\Gamma|=\langle\mathbf{k}|G_0(L, E)V$. Therefore, $|\psi\rangle$ satisfies

$$|\psi\rangle = |\mathbf{k}\rangle + [G_0(L, E)V - |\mathbf{k}\rangle\langle\mathbf{k}|G_0(L, E)V]|\psi\rangle \equiv |\mathbf{k}\rangle + \tilde{G}_0(L, E)V|\psi\rangle, \tag{2.6}$$

$$E = \epsilon(\mathbf{k}) + \langle\mathbf{k}|V|\psi\rangle. \tag{2.7}$$

equation (2.6), together with equation (2.7), is nothing but the Bethe–Goldstone equation, which has been derived with operator methods rather than for a specific system. For convenience, we define $V|\psi\rangle \equiv \hat{r}|\mathbf{k}\rangle$ and, introducing $1 = \sum_{\mathbf{k}}|\mathbf{k}\rangle\langle\mathbf{k}|$ into equations (2.6) and (2.7), we obtain

$$\langle\mathbf{k}'|\hat{r}|\mathbf{k}\rangle = \langle\mathbf{k}'|V|\mathbf{k}\rangle + \sum_{\mathbf{q}\neq\mathbf{k}}\frac{\langle\mathbf{k}'|V|\mathbf{q}\rangle\langle\mathbf{q}|\hat{r}|\mathbf{k}\rangle}{E - \epsilon(\mathbf{q})}, \tag{2.8}$$

$$E = \epsilon(\mathbf{k}) + \langle\mathbf{k}|\hat{r}|\mathbf{k}\rangle. \tag{2.9}$$

From equation (2.7), the similarity between the equation satisfied by the finite size operator $\hat{r}$ and the Lippmann–Schwinger equation for the T-matrix is apparent.

To continue, we expand the energy E following Lüscher's work. We define a dimensionless parameter δ, and write

$$E = \sum_{\nu\geqslant 0} E_\nu\delta^\nu, \tag{2.10}$$

$$E = E_0 + r(E)\delta, \tag{2.11}$$

where E_ν is unknown, and we have defined $E_0 = \epsilon(\mathbf{k})$ and $r(E) = \langle\mathbf{k}|\hat{r}(E)|\mathbf{k}\rangle$. Equation (2.11) is just equation (2.9) for $\delta = 1$, and δ shall take on the value $\delta = 1$ at the end of the calculation. We further expand

$$r(E) = \sum_{j=0}^{\infty} r_j(E - E_0)^j, \tag{2.12}$$

with $r_j = (1/j!)\partial_E^j r(E)|_{E=E_0}$. Inserting equation (2.12) into equation (2.11), and using the expansion of the energy (2.11), we obtain the first few orders as follows. Writing $E = E(\delta)$ before setting $\delta = 1$,

$$E(\delta) = E_0 + r_0\delta + r_1(E_1\delta + E_2\delta^2)\delta + r_2(E_1\delta)^2\delta + O(\delta^4). \tag{2.13}$$

We solve for E_i at each order, obtaining

$$E_1 = r_0, \tag{2.14}$$

$$E_2 = r_0 r_1, \tag{2.15}$$

$$E_3 = r_1^2 r_0 + r_0^2 r_2. \tag{2.16}$$

For a few orders in perturbation theory, the above will be sufficient.

The lowest order correction to the energy (for $\mathbf{k} = 0$) can be evaluated straight away by recalling that $\langle 0|V|0\rangle = 4\pi\hbar^2 a/mL^3$ in the Born approximation. To generalize this result to higher order in the scattering amplitude, we write equation (2.8) in operator form, that is

$$\hat{r}(E) = V + V\tilde{G}_0(L, E)\hat{r}(E). \tag{2.17}$$

To connect the Bethe–Goldstone equation with the scattering amplitudes, we should try to eliminate the interaction potential V in favour of the transition operator $T(z)$. For this, we invert the Lippmann–Schwinger equation (1.73), and expand it up to $O(T^3)$,

$$V = T(z)[1 + G_0(z)T(z)]^{-1} \tag{2.18}$$

$$= T(z) - T(z)G_0(z)T(z) + T(z)G_0(z)T(z)G_0(z)T(z) + O(T^4). \tag{2.19}$$

We define the following *formal* operator,

$$\delta\tilde{G}(L, E, z) \equiv \tilde{G}_0(E, L) - G_0(z) \equiv \delta\tilde{G}_0(L, E, z). \tag{2.20}$$

To $O(T^2)$, introducing equation (2.19) into equation (2.17), and using the above definition, we obtain

$$\hat{r}(E) = T(z) + T(z)\delta\tilde{G}_0(E, L, z)T(z) + O(T^3). \tag{2.21}$$

Note that, for equation (2.21) to be exactly correct to this order, we must assume that V has a strict finite range, that is, $V(\mathbf{r}) = 0$ for $|\mathbf{r}| > R$ for some finite $R \lesssim L/2$. Otherwise, the interaction potential cannot be probed at all interparticle distances available within the box. We shall return to this later.

A word of caution must be raised now. Note that we are seemingly adding operators belonging to (very!) different Hilbert spaces in equation (2.20) or, actually, in any relation that uses finite and infinite size quantities at the same time. We need to define matrix elements of the continuum transition and free Green's operators. In the continuum, they are well defined, that is, $\langle \mathbf{Q}'|T(z)|\mathbf{Q}\rangle$ and

$$\langle \mathbf{Q}'|G_0(z)|\mathbf{Q}\rangle = (2\pi)^3\delta(\mathbf{Q} - \mathbf{Q}')\frac{1}{z - \epsilon(\mathbf{Q})}, \tag{2.22}$$

are well defined matrix elements. The vectors $|\mathbf{Q}\rangle$ and $|\mathbf{q}\rangle$ are the same state with different normalization in the finite-L Hilbert space, that is, their position representations are identical up to a constant,

$$\langle \mathbf{r}|\mathbf{Q}\rangle = L^{3/2}\langle \mathbf{r}|\mathbf{q}\rangle = e^{i\mathbf{q}\cdot\mathbf{r}}, \tag{2.23}$$

with $\mathbf{q} = \mathbf{Q} = 2\pi\mathbf{n}_\mathbf{q}/L$, and $\mathbf{n}_\mathbf{q}$ an integer. Then, we *define* the matrix elements of the transition operator in the finite volume Hilbert space as

$$\langle \mathbf{q}'|T(z)|\mathbf{q}\rangle \equiv \frac{1}{L^3}\langle \mathbf{Q}'|T(z)|\mathbf{Q}\rangle. \tag{2.24}$$

The above relation is non-trivial since its right-hand-side has the meaning of the T-matrix in the continuum. That is, we identify the T-matrix elements in the finite volume system with the actual full, off-shell T-matrix elements in the infinite size limit. The non-interacting Green's function requires more care because it contains Dirac delta functions in the infinite size limit, see equation (2.22). To see how to sort this out, we take a look at the second-order expansion for the operator $\hat{r}(E)$, equation (2.21). Its matrix elements are given by

$$\langle \mathbf{k}'|\hat{r}(E)|\mathbf{k}\rangle = \langle \mathbf{k}'|T(z)|\mathbf{k}\rangle + \langle \mathbf{k}'|T(z)\tilde{G}_0(L, E)T(z)|\mathbf{k}\rangle - \langle \mathbf{k}'|T(z)G_0(z)T(z)|\mathbf{k}\rangle. \tag{2.25}$$

On the right-hand-side of equation (2.25) above, the first term is not a problem thanks to equation (2.24). The second term is also not a problem since we can include the finite volume resolution of the identity and all matrix elements involved are well defined. The third term is the one that is problematic in principle. We cannot use the finite volume resolution of the identity because $G_0(z)$ *is not defined within the finite volume Hilbert space*. But since matrix elements of the form $\langle \mathbf{k}'|T(z)|\mathbf{Q}\rangle$ are well defined, namely

$$\langle \mathbf{q}'|T(z)|\mathbf{Q}\rangle = \frac{1}{L^{3/2}}\langle \mathbf{Q}'|T(z)|\mathbf{Q}\rangle, \tag{2.26}$$

then we may insert the resolution of the identity in the infinite size limit, and obtain

$$\begin{aligned}\langle \mathbf{k}'|T(z)G_0(z)T(z)|\mathbf{k}\rangle &= \langle \mathbf{k}'|T(z)\int \frac{d^3Q}{(2\pi)^3}|\mathbf{Q}\rangle\langle \mathbf{Q}|G_0(z)T(z)|\mathbf{k}\rangle \\ &= \int \frac{d^3Q}{(2\pi)^3}\frac{\langle \mathbf{k}'|T(z)|\mathbf{Q}\rangle\langle \mathbf{Q}|T(z)|\mathbf{k}\rangle}{z - \epsilon(\mathbf{Q})}.\end{aligned} \tag{2.27}$$

With all of the above analysis, $r(E)$ becomes, to $O(T^2)$,

$$\langle \mathbf{k}|\hat{r}(E)|\mathbf{k}\rangle = \epsilon(\mathbf{k}) + \langle \mathbf{k}|\hat{T}(z)|\mathbf{k}\rangle + \sum_{\mathbf{q}\neq\mathbf{k}}\frac{|\langle \mathbf{k}|\hat{T}(z)|\mathbf{q}\rangle|^2}{E - \epsilon(\mathbf{q})} - \int \frac{d^3Q}{(2\pi)^3}\frac{|\langle \mathbf{k}|\hat{T}(z)|\mathbf{Q}\rangle|^2}{z - \epsilon(\mathbf{Q})}. \tag{2.28}$$

Note that, in equation (2.28), the argument z of the T-matrix is *a priori* arbitrary. This comes from the fact that equation (2.19) is exact, non-perturbatively, for all z, that is, the potential V is energy independent, and real (for complex z), if the exact expression is used. This property is lost perturbatively unless we keep order-by-order consistency in perturbation theory. We will look at this problem in a particular case later on, and show that by keeping consistency the issue of z-dependence is indeed under control.

We particularize now to the lowest scattering state, that is, we set $\mathbf{k} = 0$. We have, from equation (2.28),

$$\langle 0|\hat{r}(E)|0\rangle = \langle 0|\hat{T}(z)|0\rangle + \sum_{\mathbf{q}\neq 0}\frac{|\langle 0|\hat{T}(z)|\mathbf{q}\rangle|^2}{E-\epsilon(\mathbf{q})} - \int\frac{d^3Q}{(2\pi)^3}\frac{|\langle 0|\hat{T}(z)|\mathbf{Q}\rangle|^2}{z-\epsilon(\mathbf{Q})}. \tag{2.29}$$

A convenient choice of argument z is the actual energy. Moreover, to avoid dealing with complex quantities, we may instead replace $\hat{T}(z)$ by the reaction operator $\hat{R}(z)$, which is real for $z = E$. We finally have

$$\langle 0|\hat{r}(E)|0\rangle = \langle 0|\hat{R}(E)|0\rangle + \sum_{\mathbf{q}\neq 0}\frac{|\langle 0|\hat{R}(E)|\mathbf{q}\rangle|^2}{E-\epsilon(\mathbf{q})} - \mathrm{P}\int\frac{d^3Q}{(2\pi)^3}\frac{|\langle 0|\hat{R}(E)|\mathbf{Q}\rangle|^2}{E-\epsilon(\mathbf{Q})}, \tag{2.30}$$

where we have defined

$$\mathrm{P}\frac{1}{E-\epsilon(\mathbf{Q})} \equiv \frac{1}{2}\left[\frac{1}{E+i0^{+}-\epsilon(\mathbf{Q})} + \frac{1}{E+i0^{-}-\epsilon(\mathbf{Q})}\right]. \tag{2.31}$$

Equation (2.30) seems to suggest the off-shell (half-on-shell) amplitudes $\langle 0|\hat{R}(E)|\mathbf{q}\rangle$, that is, the short-distance details of the interaction, enter the size-dependence of the energy. This may sound natural since for small enough sizes, and for interactions that do not vanish identically beyond a certain distance (the case considered by Lüscher), the potential details have a non-vanishing effect on the energy eigenvalues. However, we shall show that the details do not matter in the large L asymptotics. For the lowest order, equation (2.14), we have

$$r_0 = \langle 0|\hat{R}(0)|0\rangle + \sum_{\mathbf{q}\neq 0}\frac{|\langle 0|\hat{R}(0)|\mathbf{q}\rangle|^2}{-\epsilon(\mathbf{q})} - \mathrm{P}\int\frac{d^3Q}{(2\pi)^3}\frac{|\langle 0|\hat{R}(0)|\mathbf{Q}\rangle|^2}{-\epsilon(\mathbf{Q})}. \tag{2.32}$$

On-shell ($\mathbf{k}'$, $\mathbf{k}$, $\mathbf{Q} \to 0$), we have

$$\lim_{\mathbf{k}',\mathbf{k},E\to 0}\langle \mathbf{k}'|\hat{R}(E)|\mathbf{k}\rangle = \frac{4\pi\hbar^2 a}{mL^3}, \tag{2.33}$$

$$\lim_{\mathbf{k},\mathbf{Q},E\to 0}\langle \mathbf{k}|\hat{R}(E)|\mathbf{Q}\rangle = \frac{4\pi\hbar^2 a}{mL^{3/2}}, \tag{2.34}$$

then we can write

$$\langle 0|\hat{R}(0)|\mathbf{q}\rangle = \frac{4\pi\hbar^2 a}{mL^3}f(\mathbf{q}), \tag{2.35}$$

$$\langle 0|\hat{R}(0)|\mathbf{Q}\rangle = \frac{4\pi\hbar^2 a}{mL^{3/2}}f(\mathbf{Q}), \tag{2.36}$$

where f is a smooth function satisfying $f(0) = 1$. The resulting expression for r_0, equation (2.32), becomes

$$r_0 = \frac{4\pi\hbar^2 a}{mL^3}\left\{1 + \frac{4\pi\hbar^2 a}{mL^3}\sum_{\mathbf{q}\neq 0}\frac{f(\mathbf{q})}{-\epsilon(\mathbf{q})} - \frac{4\pi\hbar^2 a}{m}\mathrm{P}\int\frac{\mathrm{d}^3Q}{(2\pi)^3}\frac{f(\mathbf{Q})}{-\epsilon(\mathbf{Q})}\right\}. \tag{2.37}$$

The evaluation of the large L asymptotics of r_0 is done by expanding f. This is rigorous provided the difference between the sum and integral of each individual term is convergent. Since we are here interested in the s-wave properties, we assume $f = F(q^2)$, and expand it in Taylor series as $F(q^2) = 1 + F'(0)q^2 + \cdots$. For the lowest order term, we need the following

$$\frac{1}{L^3}\sum_{\mathbf{q}\neq 0}\frac{1}{-q^2} - \mathrm{P}\int\frac{\mathrm{d}^3Q}{(2\pi)^3}\frac{1}{-Q^2} = \frac{\gamma_1}{L}, \tag{2.38}$$

where $\gamma_1 = 0.225\,785\ldots$ is obtained numerically. Since equation (2.38) is convergent, the simplest way to do this calculation is to regularize the sum and integral as

$$\frac{1}{L^3}\sum_{\mathbf{q}\neq 0}\frac{1}{-q^2} - \mathrm{P}\int\frac{\mathrm{d}^3Q}{(2\pi)^3}\frac{1}{-Q^2} = \lim_{b\to 0^+}\left[\frac{1}{L^3}\sum_{\mathbf{q}\neq 0}\frac{e^{-bq^2}}{-q^2} - \mathrm{P}\int\frac{\mathrm{d}^3Q}{(2\pi)^3}\frac{e^{-bq^2}}{-Q^2}\right]. \tag{2.39}$$

We perform the calculation with a fixed cutoff Λ for the discrete sum, and several values of b, such that equation (2.39) converges very fast. The lowest order correction in b is actually given by $-b/L^3$ (see equation (2.42) below). Therefore, for small enough b, we have

$$\frac{1}{L^3}\sum_{\mathbf{q}\neq 0}\frac{e^{-bq^2}}{-q^2} - \mathrm{P}\int\frac{\mathrm{d}^3Q}{(2\pi)^3}\frac{e^{-bq^2}}{-Q^2} = \frac{\gamma_1}{L} - \frac{b}{L^3} + \mathcal{R}(b). \tag{2.40}$$

Above, $\mathcal{R}(b)$ is the remainder. Since the next orders in the Taylor expansion do not converge and, in fact, are all divergent terms ($\propto\Lambda$ or higher powers), we can conclude that $\mathcal{R}(b)$ is not analytic in b. Indeed, if we use Poisson's summation formula, it is easy to verify that (see problem section)

$$\mathcal{R}(b) \approx 6\sqrt{\pi b}\,\frac{e^{-L^2/4b}}{L^2}, \tag{2.41}$$

which is indeed non-analytic in b, and tends to zero extremely (exponentially) fast.

The next order term can be obtained fully analytically and gives

$$\frac{1}{L^3}\sum_{\mathbf{q}\neq 0}\frac{q^2}{-q^2} - \mathrm{P}\int\frac{\mathrm{d}^3Q}{(2\pi)^3}\frac{Q^2}{-Q^2} = \frac{1}{L^3}. \tag{2.42}$$

To prove the above relation, we use Poisson's summation formula (in three dimensions), that is,

$$\frac{1}{L^3}\sum_{\mathbf{q}} 1 = \int\frac{d^3Q}{(2\pi)^3} + \sum_{\mathbf{m}\neq 0}\int\frac{d\mathbf{Q}}{(2\pi)^3}e^{-i\mathbf{m}\cdot\mathbf{Q}L}. \tag{2.43}$$

The last term on the right-hand-side of equation (2.43) vanishes, since the integral is proportional to $\delta(\mathbf{m})$. From that, equation (2.42) trivially follows.

Putting everything together, we have to this order

$$r_0 = \frac{4\pi\hbar^2 a}{mL^3}\left\{1 + c_1\frac{a}{L} + 4\pi F'(0)\frac{a}{L^3}\right\}, \tag{2.44}$$

with $c_1 = 4\pi\gamma_1 = 2.837\,297...$, which is the first numerical coefficient calculated by Lüscher, see equation (2.1). The above expression does depend on potential details at $O(L^{-6})$, in particular through $F'(0)$. We show now that this dependence is eliminated as follows. We expand the s-wave reaction matrix elements at low energy and low momenta, with the intention of going on-shell. We define

$$\langle \mathbf{K}'|R(E)|\mathbf{K}\rangle \equiv R(E, k^2), \tag{2.45}$$

where $|\mathbf{k}'| = |\mathbf{k}| = k$. The expansion takes the form, to lowest order in E and k^2,

$$R(E, k^2) = R(0, 0) + \frac{\partial R}{\partial(k^2)}(0, 0)k^2 + \frac{\partial R}{\partial E}(0, 0)E + O(k^4) + O(E^2). \tag{2.46}$$

We obviously have $R(0, 0) = 4\pi\hbar^2 a/m$, and we define two constants A and ρ such that

$$\frac{\partial R}{\partial(k^2)}(0, 0) = 2AR(0, 0), \tag{2.47}$$

$$\frac{\partial R}{\partial E}(0, 0) = \frac{m}{2\hbar^2}\rho a R(0, 0). \tag{2.48}$$

Since on-shell $E = \hbar^2 k^2/m$, using equations (2.47) and (2.48) in equation (2.46), taken on-shell, gives

$$R(E, k^2)|_{\text{on-shell}} = \frac{4\pi\hbar^2 a}{m}\left[1 + \frac{1}{2}(\rho a + 4A)k^2\right] + O(k^4). \tag{2.49}$$

We compare equation (2.49) with the effective range expansion for the reaction matrix,

$$R(E, k^2)|_{\text{on-shell}} = \frac{4\pi\hbar^2 a}{m}\left[1 + \frac{1}{2}r_e a k^2\right] + O(k^4), \tag{2.50}$$

obtaining for the effective range r_e the following value

$$r_e = \rho + \frac{4A}{a}. \tag{2.51}$$

Now recall that we defined (see equation (2.35))

$$L^3\langle 0|R(0)|\mathbf{k}\rangle = R(0, 0)[1 + F'(0)k^2] + O(k^4), \tag{2.52}$$

which implies, by comparing with equation (2.47), that $F'(0) = 2A$. Our goal, remember, is to show that only the effective range, and not other system-dependent quantities, appear in the energy. To show this, we recall (2.15) which, to leading order, is obtained from r_0 and, using equation (2.52)

$$r_1 = \frac{\mathrm{d}}{\mathrm{d}E}\langle 0|\hat{R}(E)|0\rangle|_{E=0} = \frac{1}{L^3}\frac{\partial R}{\partial E}(0, 0) = \frac{1}{L^3}2\pi\rho a^2. \tag{2.53}$$

The energy to this order is given by $E = r_0 + r_0 r_1$, so that

$$E = \frac{4\pi\hbar^2 a}{mL^3}\left[1 + c_1\frac{a}{L} + 4\pi F'(0)\frac{a}{L^3} + 2\pi\rho\frac{a^2}{L^3}\right], \tag{2.54}$$

and using the value of the effective range given in equation (2.51), together with $F'(0) = 2A$, we obtain

$$E = \frac{4\pi\hbar^2 a}{mL^3}\left[1 + c_1\frac{a}{L} + 2\pi r_e\frac{a^2}{L^3}\right]. \tag{2.55}$$

The above relation shows that the model dependence disappears by simply being consistent with the order (in a) of the expansion.

Now that the important point of model dependence has been settled, at least to this order, we improve the calculation of r_1 so that we can obtain the $O(a^3/L^5)$ correction to the energy. For this, we need to calculate the derivative with respect to the energy of $r(E)$ at the next-to-leading order. The relevant sum is given by

$$\frac{1}{L^6}\sum_{\mathbf{q}\neq 0}\frac{1}{-[\epsilon(\mathbf{q})]^2} = -\left(\frac{m}{\hbar^2}\right)^2\frac{\gamma_2}{(2\pi)^4L^2}, \tag{2.56}$$

where $\gamma_2 = 16.5323\ldots$ is a numerical constant. We also require the expansion of the potential to $O(\hat{T}^3)$ in equation (2.19), which contributes $O(a^3)$ to r_0. This contribution, which we call $r_0^{(3)}$, is given in operator form by

$$\hat{r}_0^{(3)} = \hat{T}(z)\delta\tilde{G}_0(L, E)\hat{T}(z)\delta\tilde{G}_0(L, E)\hat{T}(z). \tag{2.57}$$

To lowest order, it gives

$$r_0^{(3)} = \frac{4\pi\hbar^2 a}{mL^3}\frac{(4\pi\gamma_1)^2 a^2}{L^2}. \tag{2.58}$$

At $O(a^3/L^5)$, we take the $O(a/L^3)$ term in equation (2.44), namely the Born approximation, and multiply it with

$$\left(\frac{4\pi\hbar^2 a}{mL^3}\right)^2\sum_{\mathbf{q}\neq 0}\frac{1}{-[\epsilon(\mathbf{q})]^2}, \tag{2.59}$$

where the value of the sum above is given in equation (2.56). Therefore, we obtain for that term

$$r_0 r_1 = -\frac{4\pi\hbar^2 a}{mL^3}\left[\frac{\gamma_2}{\pi^2}\left(\frac{a}{L}\right)^2\right]. \tag{2.60}$$

We can now collect all the terms to $O(L^{-5})$, from equations (2.57) and (2.60). The resulting expression for the energy this order gives exactly equation (2.1). As a bonus, we actually obtained the leading order effective range correction, which is of $O(L^{-6})$, featured in equation (2.55).

We now show that, perturbatively, if one is consistent order-by-order, the energy z at which we evaluate the T-matrix (or R-matrix) is irrelevant, that is, the results are z-independent up until the correct order. Instead of a more general proof, we shall take a particular example, up to second order in the scattering length, which is sufficient to illustrate the problem at hand. We assume that we have a constant T-matrix, that is, it is independent of momentum. This is represented as

$$T(z) \equiv \langle \mathbf{Q}'|T(z)|\mathbf{Q}\rangle. \tag{2.61}$$

We will see in the following chapter where such a T-matrix comes from within the effective field theory paradigm. We will also simplify things by considering negative energies, that is, $z \equiv -\hbar^2\lambda^2/m$. For notational simplicity, we shall call $T(z) \equiv T(\lambda)$, and assume

$$T(\lambda) = \frac{4\pi\hbar^2}{m}\frac{1}{1/a - \lambda}. \tag{2.62}$$

Note that the above T-matrix coincides with the on-shell zero-energy T-matrix in the s-wave channel if $\lambda = 0$. Written as in equation (2.62), it represents an off-shell, low-energy T-matrix in the s-wave. We begin by expanding $T(\lambda)$ in powers of a, up to $O(a^2)$ since, as we have already mentioned, we will prove λ-independence to this order. It reads

$$T(\lambda) = \frac{4\pi\hbar^2 a}{m}[1 + \lambda a] + O(a^3). \tag{2.63}$$

The ground state energy shift, to $O(a)$, coincides with $T(\lambda)/L^3$ to that order. The energy shift at $O(a^2)$ is easily calculated using the very same approach utilized above, and reads

$$\frac{4\pi\hbar^2 a}{mL^3}\left[\lambda a + \frac{4\pi\hbar^2 a}{m}\left(\frac{\gamma_1}{L} - \frac{m}{2\pi^2\hbar^2}\int_0^\infty dQ\frac{1}{1 + (Q/\lambda)^2}\right)\right]. \tag{2.64}$$

The first and last term in the brackets of equation (2.64) add up to zero, and we recover the second-order energy shift without dependence on λ.

2.1.2 Weakly-coupled two-body scattering II: higher angular momentum

Finite-volume, weakly coupled, low-energy scattering with non-zero angular momentum is slightly more complicated than the s-wave collisions that we have studied in detail in the previous subsection. As we will see shortly, even for the ground state, degeneracy (and the lifting thereof), becomes important. Moreover, while the square of the orbital angular momentum, L^2, is a good quantum number, its azimuthal component L_z is not. This means that states which in the infinite size limit have a well defined m_z ($\hbar m_z$ being the eigenvalues of L_z), in a box become coupled with each other, in principle. One way to deal with this problem is to make (heavy) use of group theory. However, some basic notions that do not really require much background (or any) in group theory, will suffice. In particular, we will only define and make use of invariant subspaces.

Let us call $\mathcal{H}$ the Hilbert space for a non-relativistic particle (or in our case, the relative motion of two particles) in a cube of volume L^3 with periodic boundary conditions. Given a linear operator $\hat{O}$: $\mathcal{H} \to \mathcal{H}$, we say that a Hilbert subspace $\mathcal{H}'$ of $\mathcal{H}$ is an invariant subspace of $\hat{O}$ if for every $|\psi\rangle \in \mathcal{H}'$, $\hat{O}|\psi\rangle \in \mathcal{H}'$. That is, $\mathcal{H}'$ is an invariant subspace of $\hat{O}$ if it takes any state in that subspace into another state in the same subspace. We can immediately give an example of invariant subspace for any self-adjoint operator with point spectrum. Any subset $\mathcal{H}'$ of its eigenstates is indeed an invariant subspace, as the reader should be able to see immediately. For example, the kinetic energy operator for the finite volume system has infinitely many invariant subspaces consisting of all the subsets of momentum eigenstates $|\mathbf{k}\rangle$, with $\mathbf{k} = 2\pi\mathbf{n}/L$, and $\mathbf{n} \in \mathbb{Z}^3$. Our task here, as we shall see, is to find *one-dimensional invariant subspaces of the two-body interaction whose (single) elements are degenerate kinetic energy eigenstates with identical angular momentum quantum number* ℓ.

We begin by invoking the partial wave expansion for the on-shell T-matrix, equation (1.106). For orbital angular momentum quantum number ℓ, the angular dependence of the on-shell T-matrix $t(\mathbf{k}, \mathbf{k}')$ is proportional to $P_\ell(\cos\theta)$, with θ the angle between the two wave vectors $\mathbf{k}'$ and $\mathbf{k}$. We are going to rewrite the angular dependence of the partial wave projection of the interaction V in a more convenient form for our purposes for the lowest non-zero angular momentum channels, namely p- ($\ell = 1$) and d-waves ($\ell = 2$). The Legendre polynomials we need for these are given by

$$P_1(\cos\theta) = \cos\theta, \tag{2.65}$$

$$P_2(\cos\theta) = \frac{1}{2}(3\cos^2\theta - 1). \tag{2.66}$$

Since $\cos\theta = \mathbf{k}' \cdot \mathbf{k}/kk'$, then the p- and d-wave projections of the two-body interaction can be written as

$$\langle\mathbf{k}'|V_p|\mathbf{k}\rangle = F_p(k', k)\mathbf{k}' \cdot \mathbf{k}, \tag{2.67}$$

$$\langle\mathbf{k}'|V_d|\mathbf{k}\rangle = F_d(k', k)\left[(\mathbf{k}' \cdot \mathbf{k})^2 - \frac{(kk')^2}{3}\right]. \tag{2.68}$$

Generally, we have well defined values of $F_p(0, k)$ and $F_d(0, k)$, so that for zero momenta p- and d-wave scattering do not occur. This is important here because in the s-wave case we utilized $|0\rangle$ as the 'incident' wave in the Bethe–Goldstone equation, and this cannot be used as incident wave for higher angular momentum.

We find now one-dimensional invariant subspaces as explained in the previous paragraph, with the lowest possible kinetic energy. For p-waves, which have three possible values of m_z ($=-1$, 0, 1) in infinite volume, it suffices to consider the following six states, $|\pm k\hat{e}_i\rangle$, with $i = x, y, z$ and $k = 2\pi/L$, where $\hat{e}_x = (1, 0, 0)$, and so on. In fact, since p-wave scattering only occurs for antisymmetric states, we only need to consider three of these states, namely

$$|k\hat{e}_i\rangle_{\mathrm{A}} = \frac{1}{\sqrt{2}}[|k\hat{e}_i\rangle - |-k\hat{e}_i\rangle],\ i = x, y, z \tag{2.69}$$

The one-dimensional invariant subspaces of V_p are found easily. Since the space spanned by the vectors in equation (2.69) is three-dimensional, we should solve the problem (with fixed k)

$$\hat{P}_1 V_p \sum_{i=x,y,z} a_i |k\hat{e}_i\rangle_A = \lambda \sum_{i=x,y,z} a_i |k\hat{e}_i\rangle_A. \tag{2.70}$$

Above, $\hat{P}_{|\mathbf{n}|}$ is the projection operator onto the subspace of equal energy $\hbar^2(2\pi\mathbf{n}/L)^2/m$, and λ is the eigenvalue, which in this case is not relevant, as we are only interested in the eigenvectors. In this simple case, the states $|k\hat{e}_i\rangle$ ($i = x, y, z$) turn out to be the solutions to equation (2.70), which means that the three invariant subspaces S_i ($i = x, y, z$) are

$$S_i = \{|k\hat{e}_i\rangle_{\mathrm{A}}\},\ i = x, y, z. \tag{2.71}$$

We now move on to discuss d-waves. Since d-waves are symmetric, the basis of lowest energy is given by

$$|k\hat{e}_i\rangle_S = \frac{1}{\sqrt{2}}[|k\hat{e}_i\rangle + |-k\hat{e}_i\rangle],\ i = x, y, z. \tag{2.72}$$

Equation (2.70) is instead

$$\hat{P}_1 V_d \sum_{i=x,y,z} a_i |k\hat{e}_i\rangle_S = \lambda \sum_{i=x,y,z} a_i |k\hat{e}_i\rangle_S. \tag{2.73}$$

We will abuse the notation for λ and the matrix elements of V_d subsequently, since their actual values are irrelevant. The matrix elements of $\hat{P}_1 V_p \hat{P}_1$ in the symmetric basis, equation (2.73), are (proportional to)

$${}_S\langle k\hat{e}_j| V_p |k\hat{e}_i\rangle_S = \delta_{ij} - \frac{1}{3}. \tag{2.74}$$

Therefore, we need to diagonalize the following matrix

$$v_d \equiv \hat{P}_1 V_d \hat{P}_1 = \frac{1}{3}\begin{pmatrix} 2 & -1 & -1 \\ -1 & 2 & -1 \\ -1 & -1 & 2 \end{pmatrix}. \tag{2.75}$$

The eigenvalues, as we mentioned before, are irrelevant. What is not irrelevant is their degeneracy. To see that there are two degenerate eigenvalues, we calculate the determinant $\mathrm{Det}(v_d - \lambda 1)$, with v_d in equation (2.75), obtaining $-\lambda(\lambda - 3)^2/9$. The two degenerate eigenvectors with eigenvalue $\lambda = 3$ are

$$u_1^\dagger = \frac{1}{\sqrt{2}}(-1, 0, 1), \tag{2.76}$$

$$u_2^\dagger = \frac{1}{\sqrt{2}}(-1, 1, 0), \tag{2.77}$$

while the eigenvector with vanishing eigenvalue is given by

$$u_3^\dagger = \frac{1}{\sqrt{3}}(1, 1, 1). \tag{2.78}$$

Since there is one vanishing eigenvalue, this means that the vector in the Hilbert space represented by u_3, namely

$$|u_3\rangle \equiv \frac{1}{\sqrt{3}}\left[\,|k\hat{e}_x\rangle_S + |k\hat{e}_y\rangle_S + |k\hat{e}_z\rangle_S\right], \tag{2.79}$$

is not affected by d-wave interactions perturbatively, and therefore the energy shift due to these is to be calculated non-perturbatively. Finally, we have the two (perturbatively) non-trivial one-dimensional, degenerate invariant subspaces at energy $\hbar^2(2\pi/L)^2/m$, namely

$$S_1' = \{2^{-1/2}(-|k\hat{e}_x\rangle_S + |k\hat{e}_z\rangle_S)\}, \tag{2.80}$$

$$S_2' = \{2^{-1/2}(-|k\hat{e}_x\rangle_S + |k\hat{e}_y\rangle_S)\}. \tag{2.81}$$

The above invariant subspaces are somewhat inconvenient, since S_1' and S_2' are not orthogonal subspaces. But since their respective eigenvalues are identical, we can add them and subtract them at will. Hence, we can instead use the orthogonal invariant subspaces

$$S_1 = \{2^{-1/2}(-|k\hat{e}_y\rangle_S + |k\hat{e}_z\rangle_S)\}, \tag{2.82}$$

$$S_2 = \{6^{-1/2}(-2|k\hat{e}_x\rangle_S + |k\hat{e}_y\rangle_S + |k\hat{e}_z\rangle_S)\}. \tag{2.83}$$

Note that, having only two invariant subspaces at the minimal kinetic energy means that these states cannot span all possible values of m_z (=−2, −1, 0, 1, 2) for d-wave scattering.

We finally derive the 'new' version of the Bethe–Goldstone equation. To do this, let us recall equation (2.5). For the lowest states with non-zero angular momentum, we aim to avoid the sum over all states that have the same kinetic energy as the incident state, so that the Green's function can be dealt with perturbatively. Hence, let us call S_k the set of all plane-wave states (unsymmetrized) that have the same kinetic energy $\hbar^2k^2/m$. We then define the vector $|\Gamma\rangle$ as

$$\langle\Gamma| = \sum_{\mathbf{k}\in S_k} |\mathbf{k}\rangle\langle\mathbf{k}|G_0(L, E)V_\ell, \tag{2.84}$$

where V_ℓ is the projection of the two-body interaction onto the particular angular momentum channel. Clearly, $|\Gamma\rangle$ in equation (2.84) is a simple generalization of the one we defined for s-wave scattering. A trivial change of basis in equation (2.84) to that spanned by the invariant subspaces is more convenient, but we still call these vectors $|\mathbf{k}\rangle$ with a slight (but irrelevant) abuse of notation. If we take $|\psi_0\rangle$ from one of the invariant subspaces (its only member) as the incident state, and define $\hat{r}(E)|\psi_0\rangle = V|\psi\rangle$, then the Bethe–Goldstone equation reads

$$\hat{r}(E) = V + VG_0(L, E)\hat{r}(E) - V|\psi_0\rangle \sum_{\mathbf{k}\in S_k} \frac{\langle\mathbf{k}|\hat{r}(E)|\psi_0\rangle}{E - \epsilon(\mathbf{k})}, \tag{2.85}$$

$$E = \epsilon(\mathbf{k}) + \langle\psi_0|\hat{r}(E)|\psi_0\rangle, \tag{2.86}$$

$$0 = \langle\mathbf{k}|\hat{r}(E)|\psi_0\rangle,\ \mathbf{k} \in S_k\ \&\ |\mathbf{k}\rangle \neq |\psi_0\rangle. \tag{2.87}$$

We observe that the Bethe–Goldstone auxiliary condition now splits into many conditions, namely equations (2.86) and (2.87). The first condition, equation (2.86), is familiar from the previous s-wave analysis, while conditions (2.87) are the ones that triggered our anticipated analysis of invariant subspaces. Namely, the interaction, and therefore $\hat{r}(E)$, cannot couple two states for the first condition (2.86) to remain that simple. In momentum representation, equation (2.85) takes the form, for $\mathbf{k}' \notin S_k$,

$$\langle\mathbf{k}'|\hat{r}(E)|\psi_0\rangle = \langle\mathbf{k}'|V|\psi_0\rangle + \sum_{\mathbf{q}\notin S_k} \frac{\langle\mathbf{k}'|V|\mathbf{q}\rangle}{E - \epsilon(\mathbf{q})}\langle\mathbf{q}|\hat{r}(E)|\psi_0\rangle. \tag{2.88}$$

In operator form, we may write equation (2.88) as follows

$$\hat{r}(E) = V + V\tilde{G}_0^{(k)}(L, E)\hat{r}(E), \tag{2.89}$$

with notation that should be obvious to the reader by now.

Once more, we are going to solve the equations perturbatively. In this case, however, we shall explore p-wave collisions, and only to second order in the T-matrix, for illustrative purposes. To this end, we invoke equation (2.19), and write, to second order, equation (2.89) in terms of the T-matrix,

$$\hat{r}(E) = T(z) + T(z)\delta\tilde{G}_0^{(k)}(E, L, z)T(z) + O(T^3), \tag{2.90}$$

where

$$\delta\tilde{G}_0^{(k)}(E, L, z) \equiv \tilde{G}_0^{(k)}(E, L) - G_0(z). \tag{2.91}$$

The T-matrix has the same symmetry as the p-wave interaction in equation (2.67) and, therefore, we can write

$$\langle \mathbf{Q}'|T(z)|\mathbf{Q}\rangle \equiv M_p(z; Q', Q)\mathbf{Q}' \cdot \mathbf{Q}. \tag{2.92}$$

At low energies (low z) and on-shell ($Q' = Q$, $z = \hbar^2 Q^2/m + i\eta$), the function M_p characterizing the T-matrix can be written in the effective range expansion as

$$M_p(\epsilon(Q) + i\eta; Q, Q) \approx \frac{4\pi\hbar^2}{m} \frac{1}{1/a_p - \lambda_p Q^2/m} = \frac{4\pi\hbar^2 a_p}{m}\left[1 + a_p\lambda_p Q^2\right] + \cdots. \tag{2.93}$$

Above, a_p is the p-wave scattering volume, while λ_p is the effective range parameter. If we define the coupling constant $g_p = 4\pi\hbar^2 a_p/m$, and writing $k = 2\pi/L$, we may immediately calculate r_0 from equation (2.90) as

$$\frac{r_0}{k^2} = \frac{g_p}{L^3} + \frac{m}{\hbar^2}\frac{g_p^2}{3L^3}\left[\frac{1}{L^3}\sum_{\mathbf{q}\notin S_k}\frac{q^2}{k^2 - q^2} - \mathrm{P}\int\frac{d^3Q}{(2\pi)^3}\frac{Q^2}{k^2 - Q^2}\right]. \tag{2.94}$$

To lowest order, we have from the above equation

$$r_0 = \frac{16\pi^3\hbar^2 a_p}{L^5} + O(a_p^2). \tag{2.95}$$

For the next order, we will employ a bit of trickery to ease our way into extracting the correct value of the term in brackets in equation (2.94). Firstly, we can rewrite

$$\frac{1}{L^3}\sum_{\mathbf{q}\notin S_k}\frac{q^2}{k^2 - q^2} = -\frac{1}{L^3}\sum_{\mathbf{q}\notin S_k}1 + \frac{k^2}{L^3}\sum_{\mathbf{q}\notin S_k}\frac{1}{k^2 - q^2}. \tag{2.96}$$

We proceed analogously to rewrite the integral in equation (2.94). Due to the result in equation (2.42), we have

$$-\frac{1}{L^3}\sum_{\mathbf{q}\notin S_k}1 + \mathrm{P}\int\frac{d^3Q}{(2\pi)^3} = \frac{1}{L^3}\sum_{\mathbf{q}\in S_k}1 = \frac{3}{L^3}. \tag{2.97}$$

The second term in the right-hand-side of equation (2.96) (with the corresponding integral subtraction, of course), is more difficult to tackle directly, since its Poisson's summation does not converge. We may however take advantage of the fact that the cutoff version of the integral

$$\mathrm{P}\int\frac{d^3Q}{(2\pi)^2}\frac{1}{k^2 - Q^2}, \tag{2.98}$$

is energy independent (i.e., it is independent of k). Therefore, we can calculate

$$\begin{aligned}&\frac{1}{L^3}\sum_{\mathbf{q}\notin S_k}\frac{1}{k^2 - q^2} - \mathrm{P}\int\frac{d^3Q}{(2\pi)^3}\frac{1}{k^2 - Q^2}\\ &= \frac{1}{L^3}\sum_{\mathbf{q}\notin S_k}\frac{1}{k^2 - q^2} + \frac{1}{L^3}\sum_{\mathbf{q}\neq 0}\frac{1}{q^2} + \frac{\gamma_1}{L} \equiv \frac{\gamma_{p,1}}{L},\end{aligned} \tag{2.99}$$

where γ_1 is defined in equation (2.38), and given by $\gamma_1 = 0.225\,785\ldots$, while $\gamma_{p,1} = 0.030\,683\ldots$. Putting all of this together, we obtain the following to $O(g_p^2)$

$$\begin{aligned} r_0 &= \frac{16\pi^3\hbar^2 a_p}{mL^5}\left[1 + \frac{4\pi a_p}{3}\left(\frac{3}{L^3} + \left(\frac{2\pi}{L}\right)^2 \frac{\gamma_{p,1}}{L}\right)\right] + O\left(g_p^3\right) \\ &= \frac{16\pi^3\hbar^2 a_p}{mL^5}\left[1 + c_{p,1}\frac{a_p}{L^3}\right] + O\left(g_p^3\right), \end{aligned} \tag{2.100}$$

where $c_{p,1} = 17.6404\ldots$.

Now note that, in contrast to the s-wave case, r_0 in equation (2.100) above is incomplete to the order considered. This is because $r_0 = r(E_0) = r(2\pi/L)$. Therefore, it already contains effective range corrections at this level, which are of lower order than a_p^2/L^8. Indeed, using equation (2.93), the lowest effective range correction to r_0 is given by

$$\frac{64\pi^5\hbar^2\lambda_p a_p^2}{mL^7}, \tag{2.101}$$

which must be included in equation (2.100) for r_0, obtaining

$$r_0 = \frac{16\pi^3\hbar^2 a_p}{mL^5}\left[1 + c_{p,2}\lambda_p\frac{a_p}{L^2} + c_{p,1}\frac{a_p}{L^3}\right] + O\left(g_p^3\right), \tag{2.102}$$

with $c_{p,2} = 4\pi^2$. This demonstrates an instance of something quite well known—that will be explored in the following chapter—namely that the effective range is necessary for the correct description of p-wave two-body scattering, in contrast with the simpler s-wave scenario where the scattering length defines a correct, renormalizable theory.

We test now the weak-coupling s- and p-wave expansions, equations (2.1) and (2.102) with results from simple models. In particular, we choose two separable potentials in the s- and p-wave channels, respectively. The interactions, in the momentum representation, are given by

$$\langle \mathbf{Q}'|V_s|\mathbf{Q}\rangle = f(\mathbf{Q}')f(\mathbf{Q}), \tag{2.103}$$

$$\langle \mathbf{Q}'|V_p|\mathbf{Q}\rangle = f(\mathbf{Q}')f(\mathbf{Q})\mathbf{Q}'\cdot\mathbf{Q}, \tag{2.104}$$

where f is the following function

$$f(\mathbf{Q}) = g e^{-bQ^2}. \tag{2.105}$$

Note that, as such, the interactions in equations (2.103) and (2.104) are purely repulsive. The respective reaction matrices R_s and R_p can be calculated quite easily. For s-waves, we have

$$\langle \mathbf{Q}'|R_s(E)|\mathbf{Q}\rangle = A_s(E)f(\mathbf{Q}')f(\mathbf{Q}), \tag{2.106}$$

$$A_s(E) = \frac{1}{1 - I_0(E)}, \tag{2.107}$$

while for p-waves we have

$$\langle \mathbf{Q}'|R_p(E)|\mathbf{Q}\rangle = A_p(E)f(\mathbf{Q}')f(\mathbf{Q})\mathbf{Q}' \cdot \mathbf{Q}, \tag{2.108}$$

$$A_p(E) = \frac{1}{1 - \frac{1}{3}I_2(E)}, \tag{2.109}$$

where

$$I_n(E) = \mathrm{P}\int \frac{d^3Q}{(2\pi)^3}\frac{Q^n|f(\mathbf{Q})|^2}{E - \hbar^2Q^2/m}. \tag{2.110}$$

The ground state energy in a box of size L^3 with periodic boundary conditions, for bosons with interaction V_s in equation (2.103) and fermions with interaction V_p in equation (2.104), is calculated fully non-perturbatively by solving equation (2.2) numerically. The s-wave scattering length a, which is the input required in the weak-coupling expansion (2.1), is obtained from equation (2.106) as

$$\frac{4\pi\hbar^2 a}{m} = \langle 0|R_s(0)|0\rangle = A_s(0)|f(0)|^2 = g^2A_s(0) = \frac{g^2}{1 - I_0(0)}. \tag{2.111}$$

As a numerical example, we extract the scattering length a from equation (2.111) for $mbg/\hbar^2 = 0.3$, giving $a/\sqrt{b} = 0.1373...$, and compare it with a fit to the finite size ground state using equation (2.1), together with the effective range corrections (see equation (2.55)), which gives $a/\sqrt{b} = 0.1373...$ as well.

The p-wave volume a_p and effective range parameter λ_p are obtained from equation (2.108) as

$$\frac{4\pi\hbar^2 a_p}{m} = A_p(0)|f(0)|^2 = g^2A_p(0) = \frac{g^2}{1 - \frac{1}{3}I_2(0)}, \tag{2.112}$$

$$4\pi a_p^2\lambda_p = \frac{d}{dE}\left[A_p(E)\left|f\left(\sqrt{mE/\hbar^2}\right)\right|^2\right]\Bigg|_{E=0^+}. \tag{2.113}$$

It is straightforward to see that equation (2.113) gives, for λ_p, the following expression

$$\lambda_p = \frac{\hbar^2}{m}\left(-\frac{2b}{a_p} + \frac{4\pi}{3}\frac{\hbar^2 I_2'(0)}{mg^2}\right) = -\frac{\hbar^2}{m}\left(\frac{2b}{a_p} + \frac{m}{3\hbar^2}\frac{1}{\sqrt{2\pi b}}\right). \tag{2.114}$$

Fitting equation (2.102) to the numerical data for $mbg/\hbar^2 = 0.3$, we obtain $a_p/b^{3/2} = 0.476\,619...$ and $b^{1/2}\lambda_p = -4.3291$, while the exact values are $a_p/b^{3/2} = 0.476\,619...$ and $b^{1/2}\lambda_p = -4.3266....$

We have demonstrated above how weakly coupled p-wave scattering leads to an energy shift in the lowest non-interacting eigenvalue in a box of size L^3, up to second order in the p-wave scattering volume a_p. Going to higher orders is, of course, possible, but noticeably more difficult than for the s-wave problem. This is because the sums and cancellation of divergences involved are far more complicated to deal with. The situation at higher order for p-waves is replicated at lower orders for d-waves. We may continue the expansion but, fortunately, we can go on fully non-perturbatively in terms of the T-matrix, as we show in the next section.

2.1.3 Non-perturbative two-body scattering

We now generalize these results to the non-perturbative regime, leaving aside angular momentum for the moment and beginning with s-waves. Lüscher's seminal work, which took a slightly different route to the one taken here, was not capable of dealing with strong interactions, nor lower dimensions, where energy shifts are, in one dimension for example, of the same order, $O(L^{-2})$, as the non-interacting level spacing. These issues have been partially solved by working with low-energy interactions, for which the problem of obtaining finite size effects is much simpler. However, this relies on model zero-range interactions, while the method explained here is fully general.

We need to establish the Bethe–Goldstone equation (2.8), and its auxiliary condition, equation (2.9), fully in terms of the T-matrix. We will prove that

$$\hat{r}(E) = T(z) + T(z)\delta\tilde{G}_0(L, E, z)\hat{r}(E). \tag{2.115}$$

The first two terms, $O(T^2)$ and $O(T^3)$ have already been developed (see equations (2.30) and (2.57)), and indeed agree with the expansion that would result from equation (2.115) above.

The proof of equation (2.115) is as follows. Begin with an operator $\hat{s}(E)$ that satisfies equation (2.115), that is, replace $\hat{r}$ with $\hat{s}$ in the equation. We omit arguments of functions from now on to avoid overloading the notation. Use the Lippmann–Schwinger equation for the T-matrix, $T = V + VG_0T$, and insert it in the equation for $\hat{s}$, obtaining

$$\hat{s} = V + VG_0[\hat{T} + \hat{T}\delta\tilde{G}_0\hat{s}] + V\tilde{G}_0\hat{s} - VG_0\hat{s}. \tag{2.116}$$

The term in square brackets above is just $\hat{s}$, and we obtain

$$\hat{s} = V + V\tilde{G}_0\hat{s}, \tag{2.117}$$

which is the equation satisfied by $\hat{r}$, equation (2.17), and therefore $\hat{s} = \hat{r}$ and $\hat{r}$ satisfies equation (2.115), as we wanted to show. Equation (2.115) is very important, since, for potentials with finite physical range, it shows how to compute energy eigenvalues in a finite volume using the infinite volume T-matrix, without reference to the potential, in a fully non-perturbative fashion. Note also that equation (2.115) is formally valid for more than two particles.

We can recover the known result (see reference [3]) for low-energy s-wave interactions with infinite scattering length using equation (2.115). The T-matrix is purely on-shell (momentum-independent) and divergent. Therefore, equation (2.115) reduces to

$$0 = \frac{1}{L^3} + r(E)\left[\frac{1}{L^3}\sum_{\mathbf{q}\neq\mathbf{0}}\frac{1}{E - \epsilon(\mathbf{q})} - \mathrm{P}\int\frac{d^3Q}{(2\pi)^3}\frac{1}{E - \epsilon(\mathbf{Q})}\right]. \tag{2.118}$$

Note that the above equation is regular, since we have inserted the full model T-matrix, and ultraviolet subtraction schemes are not necessary (the subtraction is taken care of by the difference between the sum and the integral). The solution to the above equation is

$$E = \frac{4\pi^2\hbar^2\gamma_3}{mL^2}, \tag{2.119}$$

with $\gamma_3 = -0.0959\ldots$. We observe that the ground state energy, equation (2.119), is scale invariant, as one would expect from the unitary limit ($1/a = 0$), which has no length scales other than the size of the system.

The generalization of the above non-perturbative Bethe–Goldstone equation (2.115), to higher angular momentum channels is straightforward, and only requires changing $\delta\tilde{G}_0$ for $\delta\tilde{G}_0^{(k)}$.

Non-perturbatively, in terms of the T-matrix, the Bethe–Goldstone equation might be rather difficult to solve, unless effective field theory of interactions, discussed in the following chapter at length, is used. This is because the appearance of finite size and infinite size operators in $\delta\tilde{G}_0^{(k)}$ implies we require knowledge of (analytic continuations of) matrix elements of $\hat{r}(E)$ for all, and not only quantized values of the momentum. This is easily seen by writing, for the ground state, equation (2.115) in the momentum representation,

$$\begin{aligned}\langle\mathbf{k}'|\hat{r}(E)|0\rangle = \langle\mathbf{k}'|T(z)|0\rangle + \sum_{\mathbf{q}\neq 0}\frac{\langle\mathbf{k}'|T(z)|\mathbf{q}\rangle\langle\mathbf{q}|\hat{r}(E)|0\rangle}{E - \hbar^2q^2/m} \\ - \int\frac{d^3Q}{(2\pi)^3}\frac{\langle\mathbf{k}'|T(z)|\mathbf{Q}\rangle\langle\mathbf{Q}|\hat{r}(E)|0\rangle}{z - \hbar^2Q^2/m}.\end{aligned} \tag{2.120}$$

Without using effective field theory, there are several ways one may deal with the above issue either exactly or approximately. One example consists of setting $z = E$, using the reaction matrix instead of the T-matrix, and then using Poisson's summation formula. This approach, *a priori* exact, requires cutting off the Poisson's formula at a finite $\mathbf{m} \in \mathbb{Z}^3$, from which an approximate solution can be found. Another approximation scheme consists of the reasonable assumption that

$$\hat{r}(E) \approx \alpha R(E), \tag{2.121}$$

where $R(E)$ is the reaction operator, and α a proportionality constant. This is actually exact for separable interactions (which include effective field theory potentials). The proportionality constant can be extracted from the Bethe–Goldstone auxiliary condition and is given by

$$\alpha = \frac{EL^3}{R(E, 0)}, \tag{2.122}$$

where $R(E, 0) = \langle \mathbf{Q}' = 0|R(E)|\mathbf{Q} = 0\rangle$. Within this approximation, there is no need to solve for the Bethe–Goldstone equation and only the non-linear auxiliary condition for the energy is to be solved. A third method that yields improved results consists of replacing the integral in equation (2.120) with the sum that would result from discretizing it with periodic boundary conditions, but for length $L' > L$. In order for the quadrature points to include all points in the sum in equation (2.120), all we need is $1 < L'/L \in \mathbb{Z}$. Once more, this requires the reaction operator instead of the T-matrix, although the energy is not necessarily $z = E$. Moreover, a careful, non-trivial discretization is necessary due to the principal value integral involved, when angular momentum channels are separated unless $z = 0$. We shall not delve any deeper into these issues, since effective field theories provide a simple solution.

2.1.3.1 Long-tailed interactions and small volumes

Some of the above analysis on elastic scattering ceases to hold, or needs modification, when the interaction does not have a finite physical range. In practice, a short-range interaction in a large enough container is no different from a finite physical range interaction. However, even in large volumes, some important interactions have long tails. For instance, an r^{-6} tail occurs in atom–atom interactions (van der Waals), r^{-4} tail occurs for ion-neutral atom collisions, r^{-3} is the asymptotic behaviour of (anisotropic) dipolar interactions, while the Coulomb r^{-1} tail is present for interactions between any two charges. Power-law tails $\sim r^{-\alpha}$ ($\alpha > 0$) may vanish too slowly for the true asymptotic region of wave functions to be attained in sufficiently small boxes. Moreover, for $0 < \alpha < 3$, asymptotic scattering states do not behave in the 'usual' manner that we are familiar with using short-range potentials. Another issue we did not comment on in the previous analysis is related to the Fourier transform of the interaction potential. In the derivations, we assumed that the momentum representation of V in a finite box is formally identical to its infinite size counterpart. This is strictly true only for potentials with a finite physical range. We will begin by tackling this issue.

Consider a momentum represented potential, $\langle \mathbf{k}'|V|\mathbf{k}\rangle$, in the infinite size limit. Assume that the finite size potential in this representation, which we denote now by V_L is identical, that is $\langle \mathbf{k}'|V|\mathbf{k}\rangle = \langle \mathbf{k}'|V_L|\mathbf{k}\rangle$, which is equivalent to saying that V_L is the restriction of V to the discrete set of momenta with periodic boundary conditions. In this case, the real space representation of V_L becomes a periodic (or periodized) function. For a local potential, define $\mathbf{q} = \mathbf{k} - \mathbf{k}'$, and we have

$$V_L(\mathbf{r}) = \frac{1}{L^3}\sum_{\mathbf{q}} e^{i\mathbf{q}\cdot\mathbf{r}} V(\mathbf{q}). \tag{2.123}$$

Since the Dirac delta function admits the periodic representation $\delta_p(s) = \sum_n \exp(ins)/2\pi$ in $[-\pi, \pi]$ (mod 2π), we have

$$\delta_p(\mathbf{q}) = \frac{L^3}{(2\pi)^3} \sum_{\mathbf{n}} e^{i\mathbf{q}\cdot\mathbf{n}L}, \tag{2.124}$$

we insert $\delta_p(\mathbf{q}' - \mathbf{q})$ into equation (2.123) and obtain

$$V_L(\mathbf{r}) = \sum_{\mathbf{n}} \int \frac{d^3q'}{(2\pi)^3} e^{i\mathbf{q}'\cdot(\mathbf{r}+\mathbf{n}L)} V(\mathbf{q}') = \sum_{\mathbf{n}} V(\mathbf{r} + \mathbf{n}L), \tag{2.125}$$

which is the periodic sum of displaced copies of the infinite size potential $V(\mathbf{r})$. Clearly, if V has a finite physical range, its periodic images do not contribute to the interaction inside the box, as long as the range is smaller than the size of the system. However, for smaller boxes the potential is modified in the region of interest. This type of representation, where the momentum representation of the finite size and infinite size potentials are identical, may be suitable for *emergent* interactions. That is, when the underlying theory is more fundamental than a two-body system, and the two emergent particles have interactions that are genuinely periodic, i.e., their interactions are periodic and have continuous differentials at the edges of the box. However, if the system of interest contains truly two particles interacting via a potential, as would be the case with, say, two ^{4}He atoms in a box, periodizing the interaction is neither desirable nor necessary. Moreover, if the interactions contain a short-distance hard core, the momentum representation is to be avoided. In this case, if the box is small, further work is necessary in order to establish finite size effects. We deal now with this situation.

Assume that the interaction potential in the infinite size limit, $V(\mathbf{r})$, is spherically symmetric. This property is lost upon including it in a box if the physical range is larger than the size of the box, since we have

$$V_L(\mathbf{r}) = V(r)\theta(L/2 - |x|)\theta(L/2 - |y|)\theta(L/2 - |z|), \tag{2.126}$$

where $\theta(x)$ is the Heaviside step function. We expand V_L in spherical harmonics $Y_{\ell,m}$, obtaining

$$V_L(\mathbf{r}) = \sum_{\ell=0}^{\infty} \sum_{m=-\ell}^{\ell} V_L^{(\ell,m)}(r)\, Y_{\ell,m}(\theta, \phi), \tag{2.127}$$

$$V_L^{(\ell,m)}(r) = \int d^2\Omega\, V_L(\mathbf{r})\, Y^*_{\ell,m}(\Omega). \tag{2.128}$$

Above, Ω is the solid angle. Symmetry considerations imply that $V_L^{(\ell,m)}(r)$ is non-zero only for $m = 4n$, $n \in \mathbb{Z}$ and even ℓ. Also, $V_L^{(\ell,m)}(r) = 0$ for $r > \sqrt{3}L$, which is the maximal distance in the cube. It is easy to see that the largest contribution to $V_L(\mathbf{r})$ in equation (2.127) comes from the spherical part ($\ell = m = 0$), since for $\ell > 0$ $V_L^{(\ell,m)}(r < L) = 0$, while this is non-zero for $\ell = 0$. For $1 < r/L \leqslant \sqrt{3}$, even the spherical component of V_L gets important corrections. As a first approximation, one may use $V_L^{(0,0)}(r)$ as the effective, spherically symmetric interaction, and use all the technology developed previously in momentum space. One must take care of replacing the true, infinite size scattering parameters such as scattering length a and effective range r with the size-dependent ones associated with $V_L^{(0,\,0)}$, which we may call a_L, r_L, etc. Using $V_L^{(0,\,0)}$ instead of the entire expansion (2.127) has some problems, too. In particular, V_L is non-zero at the boundary region $\partial\mathcal{V}$ of the cube. If we define $e_0(L) = V_L(L, L, L)$, then the zero-point energy of the system is shifted to e_0, and the collision energy must be shifted by this amount, i.e., the momentum-energy relation of the system becomes $\epsilon_L(\mathbf{k}) = e_0 + \epsilon(\mathbf{k})$. Note that e_0 cannot be ignored in all instances: s-wave effective range effects come at $O(L^{-6})$, and so would e_0 corresponding to a van der Waals interaction. Obviously, this becomes more dramatic for lower powers, and becomes negligible for exponentially decaying interactions, if the length of the system is large enough.

There are two perturbative regimes we need to consider now. Firstly, the regime that is relevant for infinite physical range, where asymptotics has already been reached (large L with infinite range), and secondly, the small box regime where the short-distance interaction is probed. In the first case, one develops the analog to $\hat{r}$, $\hat{r}_L$, perturbatively in powers of the T-matrix (or R-matrix), beginning with

$$\hat{r}_L(E + e_0) = \hat{T}_L + \hat{T}_L\delta\tilde{G}_0(E + e_0)\hat{r}_L(E + e_0). \tag{2.129}$$

In the second case, it is powers of the potential that are developed, since asymptotics are never achieved, and hence we start from the equation

$$\hat{r}_L(E + e_0) = V_L + V_L\tilde{G}_0(E + e_0)\hat{r}_L(E + e_0). \tag{2.130}$$

In both equations (2.129) and (2.130), we have identified $E = \hbar^2k^2/2\mu$ with the kinetic energy in the collision, so that it is identified with the usual collision energy. The first case is formally similar to the previous analysis, with the difference that the T-matrix cannot be separated into different angular momentum channels. The second case is straightforward and results follow from usual perturbation theory.

2.2 Bound states

The energies of bound states in a finite volume are typically very well approximated by their infinite size limit. This is due to the exponentially decaying tails of their wave functions. We show now the form of these finite size corrections, following a combination of two different methods for completeness. The first method uses the momentum representation and analytic properties of amplitudes in the bound state sector, while the other method is essentially Lüscher's original derivation, which is simpler than for the scattering states. We will limit ourselves to finite physical ranges only for simplicity.

We begin by defining the finite size T-matrix $T_L(E)$, as

$$T_L(E) = V + VG_0(L, E)T_L(E). \tag{2.131}$$

The operator $T_L(E)$ has poles at every energy of the system, whether positive (scattering states) or negative (bound and some scattering states), and is real for real E. It is easy to eliminate the potential V from equation (2.131) in favour of the infinite size T-matrix, and for $E < 0$, we obtain

$$T_L(E) = T(E) + T(E)\delta G_0(L, E)T_L(E), \tag{2.132}$$

where $\delta G_0(L, E) = G_0(L, E) - G_0(E)$. The infinite size T-matrix has a pole at a bound state energy $E = E_\infty \equiv -E_B < 0$, while the finite size T-matrix has a pole that may lie near E_∞ but is displaced due to the finite size. We denote the position of the pole in a finite size by $E_L = E_\infty + \Delta$. If $|\Delta| \ll E_B$, then we can approximate

$$T(E_L) = \frac{t}{\Delta} + O(1), \tag{2.133}$$

where t is the operator-valued residue of the T-matrix at $E = E_\infty$ and, as such, is independent of energy. At the bound state energy $E = E_L$, equation (2.132) reduces to

$$t_L = \frac{t}{\Delta}\delta G_0(E_L)t_L, \tag{2.134}$$

where t_L is the residue of $T_L(E)$ at $E = E_L$, and where we have used the leading order expression for $T(E_L)$ in equation (2.133). To solve equation (2.134) approximately, we invoke equation (1.71), whose analog also holds for finite sizes. We have

$$T_L(E)G_0(L, E) = VG(L, E), \tag{2.135}$$

$$T(E)G_0(E) = VG(E). \tag{2.136}$$

Near their respective isolated poles E_L and E_∞, the Green's functions take the form

$$G(L, E) \approx \frac{|\psi_L\rangle\langle\psi_L|}{E - E_L}, \tag{2.137}$$

$$G(E) \approx \frac{|\psi\rangle\langle\psi|}{E - E_\infty}. \tag{2.138}$$

Above, $|\psi\rangle$ and $|\psi_L\rangle$ are, respectively, the normalized (this is important) bound state wave functions for infinite and finite size, associated with the eigenenergies E_∞ and E_L. Inserting (2.137) and (2.138) into equations (2.135) and (2.136), respectively, gives the following for the residues t_L and t,

$$t_L = V|\psi_L\rangle\langle\psi_L|(E_L - H_{0,L}) = (E_L - H_{0,L})|\psi_L\rangle\langle\psi_L|(E_L - H_{0,L}), \tag{2.139}$$

$$t = V|\psi\rangle\langle\psi|(E_\infty - H_0) = (E_\infty - H_0)|\psi\rangle\langle\psi|(E_\infty - H_0). \tag{2.140}$$

The above relations, inserted into equation (2.134), give us

$$\begin{aligned}\Delta\frac{E_L - \epsilon(\mathbf{k}')}{E_\infty - \epsilon(\mathbf{k}')}\psi_L(\mathbf{k}') = \psi(\mathbf{k}')\Bigg[&\sum_{\mathbf{q}}\langle\psi|(E_\infty - \epsilon(\mathbf{q})|\mathbf{q}\rangle\psi_L(\mathbf{q}) \\ &- \int\frac{d^3Q}{(2\pi)^3}(E_\infty - \epsilon(\mathbf{Q}))\psi^*(\mathbf{Q})\psi_L(\mathbf{Q})\Bigg].\end{aligned} \tag{2.141}$$

Above, we have defined $\psi(\mathbf{k}) = \langle \mathbf{k}|\psi\rangle$ and $\psi(\mathbf{Q}) = \langle \mathbf{Q}|\psi\rangle$. We perform the following approximations, which are valid to lowest order in Δ. We set E_L to E_∞, and also $\langle \mathbf{q}|\psi_L\rangle$ to $\langle \mathbf{q}|\psi\rangle$. Note that this last approximation does *not* mean $|\psi_L\rangle = |\psi\rangle$. Equation (2.141) simplifies to

$$\Delta \approx \sum_{\mathbf{q}} |\langle\psi|\mathbf{q}\rangle|^2 (E_\infty - \epsilon(\mathbf{q})) - \int \frac{d^3Q}{(2\pi)^3} |\langle\psi|\mathbf{Q}\rangle|^2 (E_\infty - \epsilon(\mathbf{Q})). \tag{2.142}$$

Now, using Poisson's summation formula, we obtain

$$\Delta \approx \sum_{\mathbf{n}\neq 0} \int \frac{d^3Q}{(2\pi)^3} |\langle\psi|\mathbf{Q}\rangle|^2 (E_\infty - \epsilon(\mathbf{Q})) e^{i\mathbf{Q}\cdot\mathbf{R}_\mathbf{n}}, \tag{2.143}$$

where $\mathbf{R}_\mathbf{n} = \mathbf{n}L$, with $\mathbf{n} \in \mathbb{Z}^3 - \{0\}$. We introduce the identity $1 = \int d^3r |\mathbf{r}\rangle\langle\mathbf{r}|$ twice in equation (2.143), obtaining, to lowest order

$$\begin{aligned} \Delta &= \sum_{\mathbf{n}\neq 0} \int d^3r \int d^3r' \psi^*(\mathbf{r}')\delta(\mathbf{R}_\mathbf{n} + \mathbf{r}' - \mathbf{r}) \left(E_\infty + \frac{\hbar^2\nabla^2}{m} \right) \psi(\mathbf{r}) \\ &= \sum_{\mathbf{n}\neq 0} \int d^3r \psi^*(\mathbf{r} - \mathbf{R}_\mathbf{n}) \left(E_\infty + \frac{\hbar^2\nabla^2}{m} \right) \psi(\mathbf{r}). \end{aligned} \tag{2.144}$$

For a non-relativistic two-body system, equation (2.144) gives the lowest order approximation to the energy shift of a bound state in a box of side L in three dimensions.

We now particularize the analysis to an s-wave bound state. Firstly, we need to recall that $|\psi\rangle$ is normalized as $\langle\psi|\psi\rangle = 1$. In the position representation, it can be written as

$$\psi(\mathbf{r}) = R_0(r) Y_{00}(\Omega), \tag{2.145}$$

where $Y_{\ell m}$ is a spherical harmonic in the usual notation, and Ω is the solid angle. Asymptotically, we define

$$R_0(r) \to \gamma \frac{e^{-\kappa r}}{r}, \quad r \to \infty \tag{2.146}$$

where γ is called the asymptotic normalization coefficient (ANC) [4]. Since $E_\infty + \hbar^2\nabla^2/m$ in equation (2.144) can be replaced with the interaction potential $V(\mathbf{r})$, and this has a finite physical range $R \ll L$, equation (2.144) becomes

$$\Delta = \frac{\gamma}{\sqrt{4\pi}} \sum_{\mathbf{n}\neq 0} \int d^3r \frac{e^{-\kappa|\mathbf{r}-\mathbf{R}_\mathbf{n}|}}{|\mathbf{r} - \mathbf{R}_\mathbf{n}|} \left(E_\infty + \frac{\hbar^2\nabla^2}{m} \right) \psi(\mathbf{r}) = -6|\gamma|^2 \frac{\hbar^2 e^{-\kappa L}}{mL} + O(e^{-\sqrt{2}\kappa L}). \tag{2.147}$$

To obtain the last equality, we have restricted the sum to the nearest points ($|\mathbf{n}| = 1$), and used $(E_\infty + \hbar^2\nabla^2/m)\exp(-\kappa r)/r = -\delta(\mathbf{r})$. Equation (2.147) explains why bound state energies converge rapidly with system size, since the energy shift is more than exponentially suppressed with system size.

2.3 Resonances

Resonant scattering can also be investigated by exploiting finite size effects. The best known way of doing this consists of considering level statistics and avoided crossings in the spectrum [5]. Although that method is well established, there are simpler ways of extracting the location and width of a resonance, at least from a theoretical viewpoint. In particular, we show how to do this by estimating the density of states in the infinite size limit using the spectrum in a finite box with twisted boundary conditions.

Density of states

We begin by reviewing the definition of density of states. This is given by the number of states at a given energy, per unit energy and volume. Given the full Hamiltonian H of the system, we define an operator $\hat{\mathcal{D}}(E)$ as

$$\hat{\mathcal{D}}(E) = \delta(E - H). \tag{2.148}$$

The imaginary part of the Green's function $G(z)$ at energy $z = E + i\eta, \eta \to 0^+$ can be written as

$$\text{Im}\, G(z) = -\lim_{\eta\to 0^+}\frac{\eta}{E - H + \eta^2} = -\pi\delta(E - H), \tag{2.149}$$

and therefore $\hat{\mathcal{D}}(E) = -\text{Im}\, G(z)/\pi$. Analogously, we have $\hat{\mathcal{D}}_0(E) \equiv \delta(E - H_0)$, and $\hat{\mathcal{D}}_0(E) = -\text{Im}\, G_0(z)/\pi$.

We use now the following relation between the T-matrix and the Green's function,

$$G(z) = G_0(z) + G_0(z)T(z)G_0(z), \tag{2.150}$$

and we find

$$\hat{\mathcal{D}}(E) = \hat{\mathcal{D}}_0(E) - \frac{1}{\pi}\,\text{Im}\,(G_0(z)T(z)G_0(z)) \equiv \hat{\mathcal{D}}_0(E) + \hat{\mathcal{D}}_I(E). \tag{2.151}$$

Let us call $|\psi_{\mathbf{k}}\rangle$ the stationary scattering state corresponding to Hamiltonian H with incident momentum $\mathbf{k}$, and $|\psi_i\rangle$ the bound states of H with energy E_i, $i = 1, 2\ldots$, if there are any. We have

$$\langle\psi_{\mathbf{k}'}|\hat{\mathcal{D}}(E)|\psi_{\mathbf{k}}\rangle = (2\pi)^3\delta(\mathbf{k} - \mathbf{k}')\delta(E - \epsilon(\mathbf{k})), \tag{2.152}$$

$$\langle\psi_{\mathbf{k}'}|\hat{\mathcal{D}}(E)|\psi_i\rangle = 0, \tag{2.153}$$

$$\langle\psi_j|\hat{\mathcal{D}}(E)|\psi_i\rangle = \delta_{ij}\delta(E - E_i). \tag{2.154}$$

We will calculate the trace of $\hat{\mathcal{D}}(E)$ in this basis. First, let us assume for a moment that all levels are discrete, i.e., bound states. Then, the trace simply becomes

$$\operatorname{Tr}\hat{D}(E) = \sum_j \langle\psi_j|\hat{\mathcal{D}}(E)|\psi_j\rangle = \sum_j \delta(E - E_i). \tag{2.155}$$

Clearly, the above is just the number of states per unit energy, and it is infinitely peaked at all bound state energies. Therefore, the density of states $\rho(E)$, in general, can be defined as

$$\rho(E) = \frac{1}{V}\operatorname{Tr}\hat{\mathcal{D}}(E). \tag{2.156}$$

The trace, when done using continuum bases, needs to be taken care of carefully. This is because continuum states are not normalized to unity, but to Dirac delta functions, as

$$\langle\psi_{\mathbf{k}'}|\psi_{\mathbf{k}}\rangle = (2\pi)^3\delta(\mathbf{k} - \mathbf{k}'). \tag{2.157}$$

To use them in a trace, we may normalize them to unity in a finite volume V. Scattering states are in one-to-one correspondence with non-interacting states. Therefore, to each non-interacting state $|\mathbf{k}\rangle$ corresponds an interacting scattering state $|\psi_{\mathbf{k}}\rangle$. That is, there is a one-to-one correspondence between state labels, and we take them as identical. With this convention (note that we could have chosen the opposite convention), in a finite volume, however, the energies of $|\mathbf{k}\rangle$ and its corresponding $|\psi_{\mathbf{k}}\rangle$ are different. We begin with one derivation that uses the non-interacting state basis $|\mathbf{k}\rangle$. We have

$$\langle\mathbf{k}'|\hat{D}_I(E)|\mathbf{k}\rangle = -\frac{1}{\pi}\langle\mathbf{k}'|\operatorname{Im}(G_0(z)T(z)G_0(z))|\mathbf{k}\rangle. \tag{2.158}$$

Near a resonance at $z = z_R = E_R - i\Gamma/2$, with E_R the resonance energy and Γ its width, $T(z)$ takes the form

$$T(z) \approx \frac{t_R}{z - z_R}, \tag{2.159}$$

with t_R energy independent, since it is a residue at a pole. We may further split $t_R = \rho_R + iI_R$, with ρ_R and I_R real operators. Equation (2.159), with $z = E + i\eta$ ($\eta \to 0^+$) can be rewritten as

$$T(z) = \frac{\left[\rho_R + iI_R\right][(E - E_R) + i\Gamma/2]}{(E - E_R)^2 + \left(\frac{\Gamma}{2}\right)^2}. \tag{2.160}$$

Therefore, introducing equation (2.160) into equation (2.158), we obtain

$$-\pi\langle\mathbf{k}|\hat{D}_I(E)|\mathbf{k}\rangle \approx \operatorname{Im}\left[\frac{1}{[z - \epsilon(\mathbf{k})]^2}\frac{\left[\rho_R(\mathbf{k}) + iI_R(\mathbf{k})\right][(E - E_R) + i\Gamma/2]}{(E - E_R)^2 + \left(\frac{\Gamma}{2}\right)^2}\right]. \tag{2.161}$$

Above, we have defined $\rho_R(\mathbf{k}) = \langle \mathbf{k}|\rho_R|\mathbf{k}\rangle$ and $I_R(\mathbf{k}) = \langle \mathbf{k}|I_R|\mathbf{k}\rangle$. To compute the trace, we shall need

$$F(E) \equiv \int \frac{d^3k}{(2\pi)^3} \langle \mathbf{k}|\hat{D}_I(E)|\mathbf{k}\rangle. \tag{2.162}$$

We need to obtain the imaginary part in equation (2.162). The full expression can be written as

$$\begin{aligned} F(E) = &\frac{1}{\pi[(E - E_R)^2 + (\Gamma/2)^2]}\Bigg[\frac{\Gamma}{2}\partial_E \mathrm{P}\int \frac{d^3k}{(2\pi)^3}\frac{\rho_R(\mathbf{k})}{E - \epsilon(\mathbf{k})} \\ &+ (E - E_R)\partial_E \mathrm{P}\int \frac{d^3k}{(2\pi)^3}\frac{I_R(\mathbf{k})}{E - \epsilon(\mathbf{k})} \\ &- \pi(E - E_R)\partial_E \int \frac{d^3k}{(2\pi)^3}\rho_R(\mathbf{k})\delta(E - \epsilon(\mathbf{k})) \\ &+ \frac{\pi\Gamma}{2}\partial_E \int \frac{d^3k}{(2\pi)^3} I_R(\mathbf{k})\delta(E - \epsilon(\mathbf{k}))\Bigg]. \end{aligned} \tag{2.163}$$

The trace in this case is obtained by making the replacement $L^{-3}\sum_{\mathbf{k}} \to \int d^3k(2\pi)^{-3}$. Since the trace is given in terms of unit-normalized states, the trace of $\hat{\mathcal{D}}_I(E)$ is given by

$$\mathrm{Tr}\, \hat{\mathcal{D}}(E) = F(E), \tag{2.164}$$

and therefore, the density of states, equation (2.156), is finally given by

$$\rho(E) = \rho_0(E) + \frac{1}{L^3}F(E), \tag{2.165}$$

with $F(E)$ given in equation (2.163).

Let us work out a simple example in which we assume that the residue t_R of the T-matrix is purely real, i.e., $t = \rho_R$. We may place cutoffs (Λ) in the momentum integrals in equation (2.164), and assume ρ_R is approximately a constant. Equation (2.164) reduces to

$$F(E) = \frac{-(E - E_R)\rho_R}{[(E - E_R)^2 + (\Gamma/2)^2]}\partial_E \int_{k<\Lambda} \frac{d^3k}{(2\pi)^3}\delta(E - \epsilon(\mathbf{k})), \tag{2.166}$$

where we have used the fact that the integral of $(E - \epsilon(\mathbf{k}))^{-1}$ is energy independent. The cutoff in the integral in equation (2.166) can be safely removed ($\Lambda \to \infty$). The integral is easily evaluated for a non-relativistic two-body system ($\epsilon(\mathbf{k}) = \hbar^2k^2/m$), as

$$\int \frac{d^3k}{(2\pi)^3}\delta(E - \epsilon(\mathbf{k})) = \frac{m}{4\pi^2\hbar^2}\sqrt{\frac{mE}{\hbar^2}}. \tag{2.167}$$

Finally, equation (2.166) reduces to

$$F(E) = \frac{-(E - E_R)\rho_R}{(E - E_R)^2 + (\Gamma/2)^2} \frac{m^{3/2}}{8\pi^2\hbar^3} E^{-1/2}. \tag{2.168}$$

We now constrain $\rho_R = \langle \mathbf{k}|\rho_R|\mathbf{k}\rangle$ using the optical theorem. On-shell, we have

$$i(T - T^\dagger) = -2\pi T^\dagger T. \tag{2.169}$$

Using equation (2.160), we have

$$2\pi i t_R^\dagger t_R = (t_R - t_R^\dagger)(E - E_R) + i\frac{\Gamma}{2}(t_R + t_R^\dagger). \tag{2.170}$$

If t_R is real, and equal to ρ_R, it simplifies to

$$\rho_R^2 = \frac{\Gamma}{2\pi}\rho_R, \tag{2.171}$$

valid *on-shell*. The expression (2.171), on-shell, can be written explicitly as

$$\int \frac{d^3q}{(2\pi)^3}\delta(\epsilon(\mathbf{k}) - \epsilon(\mathbf{q}))|\langle \mathbf{k}|\rho_R|\mathbf{q}\rangle|^2 = \frac{\Gamma}{2\pi}\langle \mathbf{k}|\rho_R|\mathbf{k}\rangle, \tag{2.172}$$

which gives

$$\rho_R = 2\pi\frac{\hbar^2\Gamma}{mk}. \tag{2.173}$$

Note, therefore, that our model with $\rho_R(\mathbf{k})$ constant is incorrect, for it violates unitarity. This is of course the case because ρ_R must be independent of energy, since it is a residue. Reworking the example, since there is only a Dirac delta function involved in the calculation, is simple, though. We obtain, using the unitarity-constrained value of ρ_R in equation (2.173), that the right-hand-side of equation (2.166) becomes

$$-\frac{(E - E_R)\Gamma}{(E - E_R)^2 + (\Gamma/2)^2}\frac{1}{4\pi E}. \tag{2.174}$$

Note that the above expression is not now $F(E)$, since the first term in the brackets in equation (2.163) may contribute to it. This term is obtained calculating the integral

$$\mathrm{P}\int_{k<\Lambda} \frac{d^3k}{(2\pi)^3}\frac{\rho_R(\mathbf{k})}{E - \epsilon(\mathbf{k})} = \frac{\hbar^2\Gamma}{\pi m}\mathrm{P}\int_0^\Lambda dk\frac{k}{E - \hbar^2k^2/m} = -\frac{\Gamma}{2\pi}\ln\left|\frac{\hbar^2\Lambda^2}{mE} - 1\right|. \tag{2.175}$$

In the limit $\Lambda \to \infty$, we have

$$\partial_E \ln\left|\frac{\hbar^2\Lambda^2}{mE} - 1\right| \to \frac{1}{E}. \tag{2.176}$$

With these results, we can finally write down the form of $F(E)$. Using equations (2.176), (2.175), and (2.174) in equation (2.163), we obtain

$$F(E) = -\frac{\Gamma}{(E - E_R)^2 + (\Gamma/2)^2} \frac{1}{4\pi E}\left[\frac{\Gamma}{\pi} + (E - E_R)\right]. \tag{2.177}$$

The above equation is the simplest possible form for $F(E)$ in a non-relativistic two-body system.

Density of states from finite size effects

So far, we have been able to relate near-resonant scattering to the density of states in the infinite size limit. This means that, strictly speaking, equation (2.164) tells us that the density of states near a resonance is still the non-interacting density of states, since $F(E)/L^3 \to 0$ as $L \to \infty$. Fortunately, we can estimate the correction $F(E)/L^3$, which contains detailed information about the resonance, equation (2.177), using the spectrum in a finite box.

To simplify the problem, we begin by considering one particle in a finite box of size L, with mass $\mu = m/2$ (the reduced mass of a system of two identical particles). What we will do here is obtain the spectrum for a variety of boundary conditions, and show that the density of states in equation (2.164) can be estimated *a priori* by performing an average over these boundary conditions. In the following, we use twisted boundary conditions, with twist angles $\theta = (\theta_x, \theta_y, \theta_z)$. The boundary conditions for position-represented states of the system $\psi(\mathbf{r})$ read

$$\psi(\mathbf{r} + \mathbf{e}_j L) = e^{i\theta_j}\psi(\mathbf{r}), \quad j = x, y, z. \tag{2.178}$$

Above, $\mathbf{e}_j$ is the unit vector in the direction j ($j = x, y, z$). Plane waves satisfying twisted boundary conditions, equation (2.178), are given by

$$\psi(\mathbf{r}) = e^{i\mathbf{k}\cdot\mathbf{r}}, \tag{2.179}$$

$$\mathbf{k} = \frac{2\pi}{L}\mathbf{n} + \frac{\theta}{L}, \tag{2.180}$$

where $\mathbf{n} = (n_x, n_y, n_z)$. For a non-relativistic system, without interactions, the spectrum is simply $\{\hbar^2\mathbf{k}^2/m\colon \mathbf{k} = 2\pi\mathbf{n}/L + \theta/L\}$. Since all states are discrete, the density of states for a given value of θ is simply

$$\rho_\theta(E) = \frac{1}{L^3}\sum_{\mathbf{k}}\delta(E - \hbar^2k^2/m), \tag{2.181}$$

with the allowed values of $\mathbf{k}$ given in equation (2.180). The density of states is, as expected, a collection of infinite peaks. Consider now the following function

$$\gamma_0(E) = \int d^3\theta\rho_\theta(E) \equiv \int_0^{2\pi} d\theta_x \int_0^{2\pi} d\theta_y \int_0^{2\pi} d\theta_z\rho_\theta(E), \tag{2.182}$$

We introduce equation (2.181) into equation (2.182), and perform the change of variable

$$\mathbf{s}_\mathbf{n} = \frac{\theta}{2\pi} + \mathbf{n}. \tag{2.183}$$

We obtain

$$\gamma_0(E) = \left(\frac{2\pi}{L}\right)^3 \sum_\mathbf{n} \int d^3 s_\mathbf{n} \delta\left(\frac{\hbar^2}{m}\left(\frac{2\pi \mathbf{s}_\mathbf{n}}{L}\right)^2 - E\right). \tag{2.184}$$

Note that, in equation (2.184), the integration limits are defined as

$$\int d^3 s_\mathbf{n} \equiv \int_{n_x}^{n_x+1} ds_x \int_{n_y}^{n_y+1} ds_y \int_{n_z}^{n_z+1} ds_z, \tag{2.185}$$

which, by definition, implies that

$$\sum_\mathbf{n} \int d^3 s_\mathbf{n} = \int_{\mathbb{R}^3} d^3 s. \tag{2.186}$$

Introducing equation (2.186) into equation (2.184), and performing the change of variables $2\pi\mathbf{s}/L = \mathbf{k}$, we obtain

$$\gamma_0(E) = \int d^3 k \delta(E - \hbar^2 k^2/m) = \rho_0(E), \tag{2.187}$$

i.e., the integral $\gamma_0(E)$ over all (vector) twist angles in a finite volume gives the infinite size non-interacting density of states $\rho_0(E)$.

The numerical procedure to make the above relations useful is as follows. We fix the system's size L and diagonalize the system N_θ times with different twisted boundary conditions $\{\theta_1, \theta_2, \dots, \theta_{N_\theta}\}$ and, for each θ_i, save the number of states $dn_E(\theta_i)$ with energies between E and $E + dE$ in bins of energy extent dE. The interacting density of states is then estimated via

$$\frac{F(E)}{L^3} \approx \frac{(2\pi)^3}{N_\theta L^3} \sum_{i=1}^{N_\theta} \frac{dn_E}{dE}(\theta_i) - \rho_0(E). \tag{2.188}$$

The major issue with using the density of states obtained by averaging over θ is the fact that twisted boundary conditions are equivalent to Bloch's theorem. The twist (vector) angle plays the role of quasi-momentum in a cubic crystal. As such, band gaps are generated at the edges of the Brillouin zone, which imply van Hove singularities and empty energy intervals in the density of states. Therefore, small volumes are required in order to be able to extract resonances in the usual non-relativistic case, so that we can resolve resonances away from van Hove singularities, especially in two and one dimensions, where the density of states diverges. At low energies, these must be far away from the first band edges, located at $\theta = (0, 0, \pi/L)$, $(0, \pi/L, 0)$ and $(\pi/L, 0, 0)$, which is rather restrictive.

The density of states and Bloch's theorem do not pose any problems for chiral states, such as those arising in topological Chern insulators. This is the case because, unlike in non-relativistic systems, chiral states do not have a time-reversal partner with opposite momentum and equal energy, and therefore a periodic potential does not open a gap.

Problems

1. Use Poisson's summation formula (see appendix A) to prove equation (2.41) from equation (2.40).
2. Starting from equation (2.131), show that the *T*-matrix in infinite space and the *T*-matrix in a finite volume can be related by equation (2.132).

References

[1] Taylor J R 2006 *Scattering Theory: The Quantum Theory of Nonrelativistic Collisions* (New York: Dover)

[2] Lüscher M 1986 Volume dependence of the energy spectrum in massive quantum field theories II. Scattering states *Commun. Math. Phys.* **105** 153

[3] Beane S R, Bedaque P F, Parreño A and Savage M J 2004 Two nucleons on a lattice *Phys. Lett.* **B585** 106

[4] Taylor J R 2000 *Scattering Theory* (New York: Dover)

[5] Döring M, Mai M and Meissner U-G 2013 Finite volume effects and quark mass dependence of the *N* (1535) and *N* (1650) *Phys. Lett.* **B722** 185

Chapter 3

Effective field theory of interactions and renormalization

Here, we introduce the notion of low-energy interactions in non-relativistic quantum mechanics. The goal is to reproduce low-energy phenomenology only, which generally generates infinities, just as in relativistic quantum field theories. The resulting low-energy theory is called effective field theory of interactions, in analogy with the non-relativistic 'pionless' effective field theory widely used in nuclear physics [1]. The predictive power of the effective theory depends upon the renormalizability of the theory, which removes the infinities while obtaining sound low-energy physics. Most of the chapter deals with perturbative and non-perturbative renormalization, which will be subsequently needed throughout the rest of the book.

3.1 Low-energy interactions

Few-body and many-body quantum systems that are treated within a realistic framework may feature detailed interaction potentials between their constituents. Besides making theoretical and computational analyses rather complicated, interaction potentials are specific to the particular system of interest. For instance, interactions between ^{4}He atoms that accurately reproduce few- and many-body phenomenology have been carefully constructed [2] and are extremely successful. However, for other systems of interest, such as ensembles of ultracold atoms, which are also manipulated by means of external fields, this approach is far too demanding, since this would have to be done for every relevant experimental situation, particular species and targeted hyperfine states, etc. This is where low-energy interactions come into play. These are universal, in the sense that they are system-independent, and only a handful of physical scattering or binding data are required for their complete characterization. Moreover, these interactions are typically far simpler than detailed potentials. The cost of this simplification is the need for regularization and renormalization, which is also studied in this chapter.

doi:10.1088/978-0-7503-3087-9ch3

The general idea behind low-energy interactions is very simple. We begin with a short-range, smooth, local, spherically symmetric two-body interaction potential $V(\mathbf{r})$, presumably realistic. In fact, we can formally consider it is the 'true' interaction potential between two particles. We can further assume that the particles are indistinguishable, say spinless bosons or spin-polarized fermions, and that the interaction is spin-independent in the latter case. We Taylor expand the interaction potential in the momentum representation, which we call $\tilde{V}(\mathbf{q})$, as

$$\tilde{V}(\mathbf{q}) = g_0 + g_2 q^2 + \cdots, \tag{3.1}$$

where g_0 and g_2 are the lowest coefficients of the Taylor expansion. Since $\mathbf{q} = \mathbf{k} - \mathbf{k}'$, with $\tilde{V}(\mathbf{q}) = \langle \mathbf{k}' | V | \mathbf{k} \rangle$, we rewrite equation (3.1) as

$$V(\mathbf{k}, \mathbf{k}') = g_0 + g_2(k^2 + k'^2) - 2g_2 \mathbf{k} \cdot \mathbf{k}' + \cdots. \tag{3.2}$$

Above, we see that the first two terms only affect s-wave scattering (they are proportional to Y_{00}), while the third term is p-wave (it only depends on $Y_{1,m}$). Therefore, we write $V(\mathbf{k}, \mathbf{k}') = V_s + V_p + \cdots$, with

$$V_s(\mathbf{k}, \mathbf{k}') = g_0 + g_2(k^2 + k'^2) + \cdots, \tag{3.3}$$

$$V_p(\mathbf{k}, \mathbf{k}') = g_1 \mathbf{k} \cdot \mathbf{k}' + \cdots, \tag{3.4}$$

where we have replaced $-2g_2$ with g_1 for simplicity. Assuming weak interactions, we can use the first Born approximation to see what these expansions reproduce, when cut at low orders. The on-shell s-wave T-matrix in the Born approximation is just $T_s^{(0)}(k) = g_0 + 2g_2k^2$, and so one expects that V_s in equation (3.3), cut at second order, will be able to reproduce scattering length and effective range effects at low energies. Later in this chapter, we will see that this holds both perturbatively and non-perturbatively, albeit with the need for highly non-trivial regularization and renormalization. Therefore, the goal of low-energy expansions of the interactions is to reproduce low-energy scattering phenomenology with simpler interaction terms.

It is also very instructive to Fourier transform these expansions back to the position representation. The s-wave interaction to leading order (LO) gives simply

$$\langle \mathbf{r}' | V_s^{\mathrm{LO}} | \mathbf{r} \rangle = g_0 \delta(\mathbf{r}) \delta(\mathbf{r} - \mathbf{r}'), \tag{3.5}$$

that is, a local ($\propto \delta(\mathbf{r} - \mathbf{r}')$), zero-range interaction. Higher-order terms lead to more singular zero-range potentials involving derivative delta functions. This, indeed, will generate short-range issues in the form of ultraviolet divergences that will be handled with regularization and renormalization.

3.2 Perturbative renormalization

3.2.1 General idea behind renormalization

In this section, we shall begin by exploring a generic example inspired by an article by Delamotte [3]. Imagine we want to calculate a certain quantity f which depends on a variable x. Among all possible methods to calculate $f(x)$, you may only be able to calculate this as a series expansion in powers of a supposedly small coupling constant g_0, upon which $f(x)$ supposedly depends, for all x. The power series may look like the following formal expression

$$f(x) = g_0 + f_2(x)g_0^2 + f_3(x)g_0^3 + \cdots = \sum_{j=1}^{\infty} f_j(x)g_0^j, \tag{3.6}$$

where we have defined $f_1(x) = 1$. The coupling constant g_0 is assumed to have no physical content, that is, it is small and arbitrary but fixed otherwise. This means that f does not really depend on g_0. A particular example with a trigonometric function will show this explicitly later. Now assume that we know a particular value of $f(x)$, say at scale μ, i.e., we know $f(\mu)$. This knowledge may come from a non-perturbative calculation, an experiment, etc. We now plug the scale μ in equation (3.6), and verify that, actually, the estimate we obtain for $f(\mu)$, perturbatively, is not good, which means, precisely, that $|f(\mu) - (g_0 + f_2(\mu)g_0^2 + \cdots)|/|f(\mu)| \gg 1$ or of order 1. Therefore, we cannot expect equation (3.6) to have any predictive value: we may not use that expression to either predict new physics or compare with experimental data. Here is where perturbative renormalization comes in; it is a way to correct the series expansion (3.6) and give it a robust predictive power. This is possible, intuitively, because knowledge of some of $f_j(x)$ and the series expansion in equation (3.6) is actually quite detailed information about the actual function $f(x)$.

Renormalization is done order-by-order. The known value of f at the scale μ is $f(\mu) \equiv g_R$, and g_R is called renormalized coupling constant. The knowledge of this value is called renormalization condition which, albeit trivial-looking, is so important that we repeat it here:

$$g_R = f(\mu). \tag{3.7}$$

To lowest order, we have $g_0 = g_R + O(g_R^2)$. This is verified directly by the renormalization condition (3.7). To the next order, we have $g_0 = g_R + \lambda_2 g_R^2$, with λ_2 a constant to be determined. Inserting this in equation (3.6) and applying the renormalization condition, we obtain $\lambda_2 = -f_2(\mu)$. Therefore, to second order in g_R, the renormalized series reads

$$f(x) = g_R + g_R^2\left[f_2(x) - f_2(\mu)\right] + O(g_R^3). \tag{3.8}$$

The third order expansion is obtained in a similar way, which the reader is encouraged to obtain in problem 1.

A rather simple example can be constructed easily. Take, for instance, the Taylor series of $\sin x$ around the origin,

$$\sin x = x - \frac{x^3}{3!} + \frac{x^5}{5!} + \cdots. \tag{3.9}$$

The sine is an entire function, and its series expansion (3.9) converges everywhere. However, maintaining a few terms only may be very inaccurate except for very small arguments x. We can use perturbative renormalization to improve the convergence rate, given only the value of the sine at a particular point μ, and the first few terms of the series. Let us define the function $f(x)$ as

$$f(x) = g_0 - \frac{g_0^2}{3!}x^2 + \frac{g_0^3}{5!}x^4 + \cdots, \tag{3.10}$$

which coincides with $\sin(x)/x$ if $g_0 = 1$. However, we do not (or are not supposed to) know this. All we know is that at some point $x = \mu$ we have equation (3.7), and that $g_R = \sin(\mu)/\mu$. We apply our perturbative renormalization scheme, and obtain to third order in the renormalized coupling constant

$$f(x) = g_R - \frac{g_R^2}{3!}(x^2 - \mu^2) + g_R^3\left[\frac{x^4}{5!} + \frac{17}{360}\mu^4 - \frac{1}{18}\mu^2 x^2\right] + O(g_R^4). \tag{3.11}$$

It is now quite intuitive to see that the above relation approximates $\sin(x)/x$, regardless of the actual value of g_0, which has been eliminated in favour of the renormalized coupling constant g_R. This is a general feature of renormalization expansions, where we say that g_0 is a *bare* coupling constant, completely unphysical. The renormalized coupling constant g_R contains relevant physical information. In this particular case, g_R contains all information about the value of f at $x = \mu$.

A more interesting case to which we will come back later has to do with contact interactions. A physically valid T-matrix can be given by $\langle \mathbf{k}'|\hat{T}(z)|\mathbf{k}\rangle = t(z)$, i.e., only dependent on energy but not on momenta. The particular case we want to illustrate has the following finite T-matrix

$$t(z) = \frac{g_R}{1 - g_R I_R(z)}, \tag{3.12}$$

where $g_R = 4\pi\hbar^2 a/m$, with a the scattering length, and

$$\begin{aligned} I_R(z) &= \int \frac{d^3k}{(2\pi)^3}\frac{z}{\epsilon(\mathbf{k})(z - \epsilon(\mathbf{k}))} \\ &= \int \frac{d^3k}{(2\pi)^3}\left[\frac{1}{z - \epsilon(\mathbf{k})} - \frac{1}{-\epsilon(\mathbf{k})}\right] \equiv I_0(z) - I_0(0). \end{aligned} \tag{3.13}$$

The very same T-matrix can be recast in the form

$$t(z) = \frac{g_0}{1 - g_0 I_0(z)}, \tag{3.14}$$

where $1/g_0 = 1/g_R + I_0(0)$. In fact, we may choose any other *bare* coupling constant g_0. The issue with this type of choice is that $I_0(z)$ is divergent, and therefore so is g_0^{-1}. Hence, while an expansion in powers of g_R in equation (3.12) is well defined and may converge (for small g_R), an expansion in powers of g_0, or any other ill-defined choice of the coupling constant, is problematic. However, this can be fixed via perturbative renormalization. Following the prescription we have outlined here, it is easy to see, order-by-order, that if we apply the renormalization condition (note that $I_R(0) = 0$)

$$t(0) = g_R, \tag{3.15}$$

the series of $t(z)$ in powers of g_0 becomes exactly the series of $t(z)$ in powers of g_R.

3.2.1.1 Several coupling constants

What if we have several coupling constants? In this case, the problem may complicate significantly, as we can see in the following example. Let us study a function f whose formal expansion can be written as

$$f(x) = g_0 + g_2 f_1^{(2)}(x) + g_0^2 f_2^{(0)}(x) + g_0 g_2 f_2^{(0,\,2)}(x) + g_2^2 f_2^{(2)}(x) + \cdots \tag{3.16}$$

Above, the functions in the expansion are generic, and may be well- or ill-defined. We assume, for simplicity, that

$$f_1^{(2)}(\mu) = 0, \tag{3.17}$$

that is, at the renormalization scale the function $f_1^{(2)}$ vanishes. The renormalization condition, equation (3.7), gives us to lowest order $g_0 = g_R$. Considering g_0 as somewhat special, we may try once more the Ansatz

$$g_0 = g_R + \lambda g_R^2 + O(g_R^3), \tag{3.18}$$

so that, to second order, the renormalization condition gives

$$0 = \left[\lambda\left(1 + g_2 f_2^{(0,\,2)}(\mu)\right) + f_2^{(0)}(\mu)\right] g_R^2 + g_R g_2 f_2^{(0,\,2)}(\mu) + g_2^2 f_2^{(2)}(\mu). \tag{3.19}$$

The above equation leaves immense freedom of choice of coupling constants *a priori*.

Let us try to find a consistent way of obtaining the parameter λ. Note that, necessarily, λ must be independent of g_R, for else the expansion (3.18) is inconsistent with perturbation theory. Therefore, the only viable options are either

$$g_R g_2 f_2^{(0,\,2)}(\mu) + g_2^2 f_2^{(2)}(\mu) = 0, \tag{3.20}$$

which implies

$$\lambda = -\frac{f_2^{(0)}(\mu)}{1 + g_2 f_2^{(0,\,2)}(\mu)}, \tag{3.21}$$

or else $g_2 = \beta g_R + O(g_R^2)$, with β some constant, independent of g_R, in which case all terms in equation (3.19) are of $O(g_R^2)$, and we have

$$\lambda + f_2^{(0)}(\mu) + \beta f_2^{(0,\,2)}(\mu) + \beta^2 f_2^{(2)}(\mu) = 0. \tag{3.22}$$

The first option, equations (3.20) and (3.21), require $g_2 \equiv 0$ or, more specifically, $g_2 f_2^{(0,\,2)}(\mu) = 0$, since λ must be independent of g_R. We shall choose the second option, that is, $g_2 = \beta g_R + O(g_R^2)$ together with equation (3.22). Clearly, we have more parameters than renormalization conditions so far, and therefore any choice of

β in equation (3.22) is possible. Obviously, we should take advantage of the further parametrization of the problem to obtain more predictive power than otherwise. Hence, we may take as a second renormalization condition $f(\mu_2) = \tilde{g}_R$. For μ_2 sufficiently close to μ, we may simply replace the second renormalization condition for a condition for the derivative, which may be simpler. This reads

$$f'(\mu) = \alpha_R, \tag{3.23}$$

where we have defined a second renormalized coupling constant α_R. We will also assume that $f_1^{(2)'}(x) = 0$ in equation (3.16), for simplicity. We differentiate equation (3.16) and insert the expansions of g_0 and g_2 in terms of g_R into it. Defining further

$$\gamma_2^{(0)} = \partial_x f_2^{(0)}(x)|_\mu, \tag{3.24}$$

$$\gamma_2^{(0,\,2)} = \partial_x f_2^{(0,\,2)}(x)|_\mu, \tag{3.25}$$

$$\gamma_2^{(2)} = \partial_x f_2^{(2)}(x)|_\mu, \tag{3.26}$$

we obtain $\alpha_R = \sigma_R g_R^2$, with

$$\sigma_R = \gamma_2^{(0)} + \beta\gamma_2^{(0,\,2)} + \beta^2\gamma_2^{(2)}, \tag{3.27}$$

which is the equation to be satisfied by β. Inserting β from equation (3.27) into equation (3.22), we obtain the value of λ. Finally, we can obtain the renormalized series that estimates $f(x)$, which reads

$$\begin{aligned} f(x) = g_R + g_R^2\Big\{ & f_2^{(0)}(x) - f_2^{(0)}(\mu) + \beta\big[f_2^{(0,\,2)}(x) - f_2^{(0,\,2)}(\mu)\big] \\ & + \beta^2\big[f_2^{(2)}(x) - f_2^{(2)}(\mu)\big]\Big\}. \end{aligned} \tag{3.28}$$

Clearly, equation (3.28) tells us that to second order in the coupling constant g_R, the usual subtraction at scale μ (corresponding to $\beta = 0$) is done for every one of the expansion functions involved. The value of β is given by the second renormalization condition, namely $f'(\mu) = \sigma_R g_R^2$ together with equation (3.27). Note that, since we are requiring order-by-order consistency in powers of g_R, the expansion in equation (3.28) is a weak-coupling expansion valid for small g_R. It does depend on $f'(\mu)$ via β and its dependence on σ_R, but this dependence is not necessarily analytic in σ_R, as we only require it for the renormalized coupling constant g_R. In other words, we have turned an expansion in two bare coupling constants g_0 and g_2, equation (3.16), into an expansion in only one renormalized coupling constant, g_R, which depends non-perturbatively on the ratio, σ_R, between $f'(\mu)$ and g_R^2.

The case studied above, in which $f_1^{(2)}(\mu) = f_1^{(2)'}(\mu) = 0$ is quite specific. In particular, it can be used for the example of a theory that reproduces scattering length and effective range correctly, provided $\mu = 0$. However, a more general scenario may not fulfil these conditions. Therefore, we proceed to study the renormalization of equation (3.16) in the case $f_1^{(2)}(\mu) \neq 0$ and $f_1^{(2)'}(\mu) \neq 0$. The two renormalization conditions read, as before, $f(\mu) = g_R$ and $f(\mu_2) = \tilde{g}_R$. To leading order, the first condition implies that

$$g_2 = \frac{g_R - g_0}{f_1^{(2)}(\mu)}. \tag{3.29}$$

We need the second renormalization condition at μ_2 as well. Using also equation (3.29), and further defining

$$c = \frac{f_1^{(2)}(\mu_2)}{f_1^{(2)}(\mu)}, \tag{3.30}$$

$$\frac{\tilde{g}_R}{g_R} = \beta, \tag{3.31}$$

$$\lambda = \frac{\beta - c}{1 - c}, \tag{3.32}$$

$$\tilde{\lambda} = \frac{1 - \lambda}{f_1^{(2)}(\mu)}, \tag{3.33}$$

we obtain, to leading order,

$$g_0 = \lambda g_R, \tag{3.34}$$

$$g_2 = \tilde{\lambda} g_R. \tag{3.35}$$

Hence, we can consider the following expansion (with λ and $\tilde{\lambda}$ already fixed) for the bare coupling constants g_0 and g_2:

$$g_0 = \lambda g_R + \lambda_2 g_R^2 + O(g_R^3), \tag{3.36}$$

$$g_2 = \tilde{\lambda} g_R + \tilde{\lambda}_2 g_R^2 + O(g_R^3), \tag{3.37}$$

where λ_2 and $\tilde{\lambda}_2$ two constants to be determined using the renormalization conditions. Using these, to second order in g_R, we readily obtain

$$\tilde{\lambda}_2 = -\frac{\lambda^2 \Delta_2^{(0)} + \lambda\tilde{\lambda}\Delta_2^{(0,2)} + \tilde{\lambda}^2 \Delta_2^{(2)}}{\Delta_1^{(2)}}, \tag{3.38}$$

$$\lambda_2 = -\left[\tilde{\lambda}_2 f_1^{(2)}(\mu) + \lambda^2 f_2^{(0)}(\mu) + \lambda\tilde{\lambda} f_2^{(0,2)}(\mu) + \tilde{\lambda}^2 f_2^{(2)}(\mu)\right]. \tag{3.39}$$

Above, we have defined $\Delta_1^{(0)} = f_1^{(2)}(\mu_2) - f_1^{(2)}(\mu)$ and analogously for the other Δ parameters.

Finally, if instead of defining $f(\mu_2) = g_R$ we took $f'(\mu) = \alpha_R$, then we would obtain, after defining $\alpha_R = \gamma g_R$,

$$\lambda = 1 - \gamma \frac{f_1^{(2)}(\mu)}{\Delta_1^{(2)}(\mu)}, \tag{3.40}$$

$$\tilde{\lambda} = \frac{\gamma}{\Delta_1^{(2)}(\mu)}, \tag{3.41}$$

where we have defined $\Delta_1^{(2)}(\mu) = f_1^{(2)\prime}(\mu)$, and analogously for the rest of the functions. The rest of the result, namely equations (3.36) and (3.38), as well as the values for the coefficients in equations (3.38) and (3.39), remain identical, except for the replacement of $\Delta_2^{(0)}$ with $\Delta_2^{(0)}(\mu)$, and analogously for the remaining functions.

3.2.2 A theorem by Riemann

The reader may have noticed how, very deliberately, we have avoided an introduction to perturbative renormalization with the excuse of infinite quantities that must be renormalized in order for the physical result to be finite. The fact is that the physical theory must give finite results from the start, and our poor choice of 'coupling constant' (the bare coupling constant) is the reason for poor convergence, or even infinities in perturbation theory. As was shown in the previous section, perturbative renormalization does not depend on infinities at all, and can as well be performed on finite series with either good or bad convergence properties. To further develop these ideas from a more theoretical perspective, we outline here a theorem by Riemann. This will be important to show, in the next section, that perturbative renormalization is just a conformal rearrangement of the perturbation series.

The statement of Riemann's theorem below follows reference [4]. Consider a complex function A of a complex variable λ in the disk of radius $|g_0|$, $C(|g_0|)$, with $g_0 \in \mathbb{C}$. Assume A is holomorphic in $C(|g_0|)$. Consider a simply connected region $D(g_0)$ such that $0 \in D(g_0)$ and $g_0 \in \partial D(g_0)$, and such that A is holomorphic in $D(g_0)$. Then, there exists a unique bijective conformal map $\Lambda(z; g_0)$ of the unit circle onto $D(g_0)$ such that its centre $z = 0$ maps into 0, and $z = 1$ maps into g_0, that is

- $\Lambda(z; g_0)$ is holomorphic in the unit disk ($|z| \leqslant 1$)
- $\Lambda(0; g_0) = 0$ and $\Lambda(1; g_0) = g_0$.

Riemann's theorem can be used to rearrange series expansions rather simply. Since $\Lambda(1; g_0) = g_0$, then the function A of the complex variable g_0 can be written as $A(g_0) = A(\Lambda(1; g_0))$. Therefore, we can write

$$A(\Lambda(z; g_0)) = \sum_{j=0}^{\infty} U_j(g_0) z^j, \tag{3.42}$$

$$A(g_0) = \sum_{j=0}^{\infty} U_j(g_0). \tag{3.43}$$

The coefficients $U_j(g_0)$ in the series expansion are easily obtained using Leibniz's rule. The first few are given by

$$U_0(g_0) = A_0, \tag{3.44}$$

$$U_1(g_0) = A_1\Lambda_1(g_0), \tag{3.45}$$

$$U_2(g_0) = A_2\Lambda_1(g_0) + A_1\Lambda_2(g_0), \tag{3.46}$$

where $A_j = \partial_\lambda^j A(\lambda)|_{\lambda=0}/j!$ and $\Lambda_j = \partial_z^j \Lambda|_{z=0}/j!$.

3.2.3 Perturbative renormalization as a conformal rearrangement

The background material just introduced should already tell a story about what we are going to prove now. Namely, that perturbative renormalization is nothing but a conformal rearrangement of an ill-defined series expansion. Since we have all the technology already in place, we show this order-by-order up to and including third order.

Take the function in equation (3.6). The bare coupling constant is, as we have insisted already, not a physically relevant quantity. We know the value of the function f at the scale $x = \mu$, which is the renormalization scale. This value of the function gives us the renormalization condition $f(\mu) = g_R$. From the perspective of Riemann's conformal mapping theorem, the variable x is a fixed parameter. The order-by- order renormalization prescription of section 3.2.1 tells us that the two series

$$f(x) = g_0 + f_2(x)g_0^2 + \cdots \tag{3.47}$$

$$f(x) = g_R + f_{2,R}(x)g_R^2 + \cdots, \tag{3.48}$$

are related via a series expansion of g_0 in powers of g_R. Using the results of the previous section, we identify

$$\Lambda(z;\, g_0) = zg_R + z^2g_R^2\left[f_{2,R}(x) - f_2(x)\right] + O(z^3). \tag{3.49}$$

Since $\Lambda(1;\, g_0) = g_0$, we have that, to this order,

$$g_0 = g_R + g_R^2\left[f_{2,R}(x) - f_2(x)\right]. \tag{3.50}$$

Applying the renormalization condition to equation (3.48), we have $f_{2,R}(\mu) = 0$. If, moreover, we *impose* that the bare coupling constant does not depend on x, then it must be the same for all values of x, in particular for $x = \mu$, which implies that, to this order,

$$g_0 = g_R - f_2(\mu)g_R^2, \tag{3.51}$$

which gives an implicit (although in this case it can be made explicit) relation $g_R = g_R(g_0)$. This means that the conformal mapping in equation (3.49) takes the form

$$\Lambda(z;\, g_0) = zg_R(g_0) - z^2f_2(\mu)[g_R(g_0)]^2 + O(z^3). \tag{3.52}$$

The next order can be obtained easily as well. Again, we have

$$f(x) = g_0 + f_2(x)g_0^2 + f_3(x)g_0^3 + \cdots \tag{3.53}$$

$$f(x) = g_R + f_{2,R}(x)g_R^2 + f_{3,R}(x)g_R^3 + \cdots, \tag{3.54}$$

which immediately gives

$$\begin{aligned}\Lambda(z;\, g_0) = {} & zg_R + z^2g_R^2\big[f_{2,R}(x) - f_2(x)\big] \\ & + z^3g_R^3\Big[f_{3,R}(x) - f_3(x) - 2f_2(x)\big(f_{2,R}(x) - f_2(x)\big)\Big] + O(z^4).\end{aligned} \tag{3.55}$$

The renormalization condition, if applied order-by-order, gives $f_{2,R}(\mu) = 0$ and $f_{3,R}(\mu) = 0$, and using $\Lambda(1;\, g_0) = g_0$, we have

$$g_0 = g_R - f_2(\mu)g_R^2 + g_R^3\big[2(f_2(\mu))^2 - f_3(\mu)\big], \tag{3.56}$$

assuming that g_0 does not depend on x.

This procedure can be continued to any desired order, and proves that the perturbative renormalization introduced here is exactly equivalent to a conformal rearrangement of the perturbative series in the spirit of Riemann's conformal mapping theorem.

3.2.4 Perturbative renormalization of operator series

The formalism developed so far only deals with functions of a complex variable. These, as we will see later, may correspond with the Born series for an on-shell scattering amplitude. However, successful renormalization of a full theory requires the renormalization of operator-valued functions. These are usually infinite-dimensional, but one may also think (in private) of matrices, with a finite number of degrees of freedom.

We begin by defining an operator $\hat{O}(z)$, which is holomorphic in the complex variable z. The operator is expanded in powers of another operator $\hat{W}_0$, assumed (whether correctly or incorrectly) to be 'small'. In this expansion, other operators may be involved, due to the non-commutativity, in general, of different operators. We assume for the time being that there is one such operator, which depends generally on z, which we denote $\hat{\Omega}(z)$. The series may look like

$$\hat{O}(z) = \hat{W}_0 + \lambda_{2,\,0}\hat{W}_0^2\hat{\Omega}(z) + \lambda_{1,\,1}\hat{W}_0\hat{\Omega}(z)\hat{W}_0 + \lambda_{0,\,2}\hat{\Omega}(z)\hat{W}_0^2 + \cdots. \tag{3.57}$$

Note how, above, the operators are allowed to appear in any non-equivalent order, and $\lambda_{i,j}$ are constants. We can call the operator $\hat{W}_0$ the *bare* coupling operator, in analogy with the renormalization of standard functions. The renormalization condition at scale $z = \mu$, in this case, is an operator condition, and we state it as

$$\hat{O}(\mu) = \hat{W}_R, \tag{3.58}$$

where $\hat{W}_R$ can be called the *renormalized* coupling operator. Analogously to the renormalization of functions, we expand $\hat{W}_0$ in powers of the renormalized coupling $\hat{W}_R$ as

$$\hat{W}_0 = \hat{W}_R + \gamma_{2,0}\hat{W}_R^2\hat{\Gamma} + \gamma_{1,1}\hat{W}_R\hat{\Gamma}\hat{W}_R + \gamma_{0,2}\hat{\Gamma}\hat{W}_R^2. \tag{3.59}$$

Applying the renormalization condition, we immediately obtain

$$\gamma_{ij} = -\lambda_{ij}, \tag{3.60}$$

$$\hat{\Gamma} = \hat{\Omega}(\mu), \tag{3.61}$$

which results, to this order, in the following renormalized series for $\hat{O}(z)$,

$$\begin{aligned}\hat{O}(z) = \hat{W}_R + \gamma_{2,0}\hat{W}_R^2[\hat{\Omega}(z) - \hat{\Omega}(\mu)] + \gamma_{1,1}\hat{W}_R[\hat{\Omega}(z) - \hat{\Omega}(\mu)]\\ \hat{W}_R + \gamma_{0,2}[\hat{\Omega}(z) - \hat{\Omega}(\mu)]\hat{W}_R^2.\end{aligned} \tag{3.62}$$

This type of expansion is quite common in the Born series for the T-matrix. More generally, the operator $\hat{O}$ may be the solution to an integral equation with kernel $\hat{K}(z)$ of the form

$$\hat{O}(z) = \hat{W}_0 + \hat{K}(z)\hat{O}(z). \tag{3.63}$$

In the case of scattering, $\hat{K}(z)$ depends explicitly on $\hat{W}_0$, but this is not necessarily the case. This still allows for perturbative renormalization, even though *a priori* the solution to the integral equation may be linear in $\hat{W}_R$ if $\hat{K}(z)$ does not depend on $\hat{W}_0$.

3.2.5 Renormalization of singular Born series

Now that we have all the perturbative renormalization theory in place, we are ready to apply it to real problems. In non-relativistic quantum mechanics, the most relevant case in which one may need to deal with perturbative renormalization is in perturbative scattering theory. Perturbation theory is systematized by means of the Born series. The T-matrix in a two-body problem, after separation of centre of mass and relative momenta, is expanded as

$$\hat{T}(z) = V + VG_0(z)V + VG_0(z)VG_0(z)V + \cdots = V\sum_{n=0}^{\infty}[G_0(z)V]^n. \tag{3.64}$$

Typically, one requires V to be small, in the sense of an operator norm, for the series above to converge. If all the terms in equation (3.64) are finite, as if, for example, V denotes the Yukawa potential, one may still apply perturbative renormalization, if we knew the *full* T-matrix at a particular energy $z = \mu + i0^+$, to improve the Born series at energies close to μ. However, this is evidently not the usual case since knowing the full T-matrix means we know $\langle \mathbf{k}'|\hat{T}(\mu + i0^+)|\mathbf{k}\rangle$ for all $\mathbf{k}'$ and $\mathbf{k}$, that is, both on- and off-shell. Perturbative renormalization is useful when using low-energy interactions, which are singular and yield singular Born series with ultraviolet divergences. Since the T-matrices for low-energy interactions are typically very simple, only fulfilling basic physical properties such as unitarity, and are essentially only needed on-shell, we can easily renormalize the Born series and give it predictive power.

We begin with the simplest of all low-energy interactions. This is given by a contact interaction, whose *bare* potential is a constant in momentum space,

$$\langle \mathbf{k}'|V|\mathbf{k}\rangle = g_0. \tag{3.65}$$

This interaction is singular, as is observed by calculating the second Born approximation in equation (3.64). In particular, the term $VG_0(z)V$ is infinite. To see this, we obtain its matrix elements as

$$\langle \mathbf{k}'|VG_0(z)V|\mathbf{k}\rangle = \int \frac{d^3q}{(2\pi)^3}\frac{\langle \mathbf{k}'|V|\mathbf{q}\rangle\langle \mathbf{q}|V|\mathbf{k}\rangle}{z - \epsilon(\mathbf{q})} = \frac{g_0^2}{2\pi^2}\int_0^\infty dq\frac{q^2}{z - \hbar^2q^2/m}, \tag{3.66}$$

which is clearly divergent. Because the interaction is a constant in the momentum representation, the Born series, equation (3.64), only produces constant terms. We insist, once more, that constant in the context of scattering theory means that it does not depend on momentum, yet a constant depends on energy (z). This implies that the renormalization condition, which is an operator relation, equation (3.58), can be chosen to be a constant in the momentum representation, that is

$$\hat{T}(\mu) = \hat{W}_R, \tag{3.67}$$

is written in the momentum representation as

$$\langle \mathbf{k}'|\hat{T}(\mu)|\mathbf{k}\rangle = g_R, \tag{3.68}$$

where g_R is the renormalized coupling constant at scale μ. For this lowest-order interaction, the renormalization scale is typically chosen to be zero, in which case we have

$$g_R = \frac{4\pi\hbar^2 a}{m}, \tag{3.69}$$

where a is the s-wave scattering length. We can now proceed to obtain the renormalized Born series order-by-order, using the results of the previous subsections. For simplicity in the notation, we define $\langle \mathbf{k}'|\hat{T}(z)|\mathbf{k}\rangle \equiv t(z)$, since it is a constant. To lowest order, we have $t(z) = g_R + O(g_R^2)$. To second order in g_R, we have

$$t(z) = g_R + \frac{g_R^2}{2\pi^2}\int_0^\infty dq\frac{mz/\hbar^2}{z - \hbar^2q^2/m} + O(g_R^3), \tag{3.70}$$

where we have used $\mu = 0$ as the renormalization scale. Note how equation (3.70) is finite, since for large q the integrand vanishes as q^{-2}. Therefore, we have already accomplished renormalization to second order in the coupling constant g_R. Since this depends on a measurable quantity only, namely the scattering length, we already have a theory in equation (3.70) with predictive power, as promised. In particular, it gives the low-energy behaviour of the scattering amplitude for small scattering lengths. We can continue this procedure to arbitrary order, but shall stop here—all orders will be obtained non-perturbatively in the next section.

The first example above shows the power of perturbative renormalization in a very simple context. Things get harder if more realistic scenarios are considered. The next-to-leading order effects at low energies and in the s-wave sector are controlled by the effective range r. The on-shell T-matrix, in the effective range expansion including up to the effective range only, is given by

$$\langle \mathbf{k}'|\hat{T}(z)|\mathbf{k}\rangle = \frac{4\pi\hbar^2/m}{1/a - rk^2/2 + O(k^4) + ik}, \tag{3.71}$$

where we have used the on-shell conditions $|\mathbf{k}'| = |\mathbf{k}| = k$ and $z = \hbar^2k^2/m + i0^+$. For the renormalization procedure, we shall only need the quantity $\mathrm{Re}(1/T^{\mathrm{on}}(k))$, defined as

$$\frac{4\pi\hbar^2}{m}\mathrm{Re}\left(\frac{1}{T^{\mathrm{on}}(k)}\right) \equiv \frac{1}{a} - \frac{1}{2}rk^2 + O(k^4). \tag{3.72}$$

The imaginary part of the inverse on-shell T-matrix is irrelevant (but not unimportant, as unitarity must *always* be checked for consistency), since it is fixed by unitarity. The physically measurable quantity here is the on-shell amplitude. Therefore, there is a great deal of freedom in the choice of the full T-matrix, since many different off-shell T-matrices coincide on-shell. However, low-energy singular interactions very clearly show which types of T-matrices one obtains at low energies and their regimes of validity. Since we aim at fixing two parameters of the theory (a and r), we will need two bare coupling constants, g_0 and g_2. As before, the (operator) renormalization condition at scale μ is just

$$\hat{T}(\mu) = \hat{W}_R. \tag{3.73}$$

The bare low-energy interaction is given by (see equation (3.4))

$$\langle \mathbf{k}'|V|\mathbf{k}\rangle = g_0 + g_2(k'^2 + k^2). \tag{3.74}$$

Let us work out the renormalization of the T-matrix for the naïve interaction (3.74). To second order, we have

$$\hat{T}(z) = \hat{W}_R + \hat{W}_R[G_0(z) - G_0(\mu)]\hat{W}_R. \tag{3.75}$$

So far, we have not given a precise form to the matrix elements of $\hat{T}(\mu)$. In the momentum representation, they read

$$\langle \mathbf{k}'|\hat{T}(z)|\mathbf{k}\rangle = g_{0,R} + g_{2,R}(k'^2 + k^2) + A + B(k'^2 + k^2) + Ck'^2k^2, \tag{3.76}$$

where we have defined renormalized coupling constants $g_{0,R}$ and $g_{2,R}$, whose physical content is yet unspecified, and three constants (which do depend on z):

$$A = \int \frac{d^3q}{(2\pi)^3} G_R(z, \mu; \mathbf{q})\left[g_{0,R}^2 + 2g_{0,R}g_{2,R}q^2 + g_{2,R}q^4\right] \tag{3.77}$$

$$B = \int \frac{d^3q}{(2\pi)^3} G_R(z, \mu; \mathbf{q})\left[g_{0,R}g_{2,R} + q^2 g_{2,R}^2\right] \tag{3.78}$$

$$C = \int \frac{d^3q}{(2\pi)^3} G_R(z, \mu; \mathbf{q}) g_{2,R}^2. \tag{3.79}$$

Note that A and B are divergent. We see, therefore, that care must be taken in the low-energy expansion of the bare interaction. In particular, regularization in some form must be imposed before the calculation is performed. In this case, it suffices to consider, instead of equation (3.74), the following bare interaction

$$\langle \mathbf{k}'|V|\mathbf{k}\rangle = g_0 + g_2(k'^2 + k^2)\theta(\Lambda - k)\theta(\Lambda - k'), \tag{3.80}$$

where Λ clearly acts now as a cutoff for the momentum in the corresponding integrals and the low-energy expansion. Note how, in equation (3.80), the cutoffs are only applied to the second term, but are not necessary for the first one. The cutoff Λ is to be taken to infinity at the end of the calculation, since it carries no physical meaning. Since we are fixing scattering length and effective range, the perturbative on-shell T-matrix in equation (3.76) cannot contain the term $\propto k'^2k^2$. Therefore, the constant C must vanish. Using the cutoff version of the interaction, equation (3.80), we obtain

$$C = \lim_{\Lambda\to\infty} g_{2,R}^2 \int_0^\Lambda \frac{dqq^2}{2\pi^2} \frac{\mu - z}{(z - \hbar^2q^2/m)(\mu - \hbar^2q^2/m)}. \tag{3.81}$$

We have $C = 0$ in equation (3.81), as we had required, provided that $g_{2,R} \to 0$ as $\Lambda \to \infty$. For the constant B, we have

$$B = \lim_{\Lambda\to\infty}\left[g_{0,R}g_{2,R} \int_\Lambda \frac{d^3q}{(2\pi)^3} G_R(z, \mu; \mathbf{q}) + g_{2,R}^2 \int_\Lambda \frac{d^3q}{(2\pi)^3} q^2 G_R(z, \mu; \mathbf{q})\right]. \tag{3.82}$$

Because of the condition for C to vanish, the first term on the right-hand-side of equation (3.82) vanishes, too, because it is easy to see that $g_{0,R}$ is finite. Therefore, equation (3.82) simplifies to

$$B = \lim_{\Lambda\to\infty} g_{2,R}^2 \int_\Lambda \frac{d^3q}{(2\pi)^3} q^2 G_R(z, \mu; \mathbf{q}). \tag{3.83}$$

The constant A is calculated as

$$A = \lim_{\Lambda\to\infty} \int_\Lambda \frac{d^3q}{(2\pi)^3} G_R(z, \mu; \mathbf{q})\left[g_{0,R}^2 + 2g_{0,R}g_{2,R}q^2 + g_{2,R}^2 q^4\right]. \tag{3.84}$$

We need A to be finite. The second and third terms together need to yield a finite result. We define a finite constant λ_R such that, to $O(z)$, the last two terms in equation (3.84) yield $\lambda_R z$. Doing the integrals with a finite cutoff gives

$$\lambda_R = -\lim_{\Lambda\to\infty}\left(\frac{m}{\hbar^2}\right)^2 \frac{\Lambda}{2\pi^2} g_{2R}\left[2g_{0,R} + g_{2,R}\frac{\Lambda^2}{3}\right]. \tag{3.85}$$

The above relation gives $g_{2,R}$ as a function of λ_R for large cutoffs (i.e., before taking the limit). It is readily checked that, as a consequence, $\lambda_R < 0$, and $g_{2,R} \sim \Lambda^{-3/2}$ for large values of the cutoff. Expanding the T-matrix on-shell ($z = \hbar^2k^2/m$), we obtain the relation between the physical parameters and the coupling constants as

$$g_{0,R} = \frac{4\pi\hbar^2 a}{m}, \tag{3.86}$$

$$\lambda_R = 2\pi a^2 r. \tag{3.87}$$

From equation (3.87), we see that $\lambda_R < 0$ implies that the effective theory requires $r < 0$ in order for it to be renormalizable perturbatively. The bound on the effective range holds non-perturbatively, as we will see in the next section, and is a consequence of Wigner's bound [5].

We now turn our attention to low-energy p-wave interactions. These are more involved than their s-wave counterpart since, as we will see, the p-wave range cannot be set to zero to obtain a renormalizable theory. We begin directly by including the LO and NLO bare interaction terms

$$\langle \mathbf{k}'|V|\mathbf{k}\rangle = [g_1 + g_3(k'^2 + k^2)]\mathbf{k}' \cdot \mathbf{k}. \tag{3.88}$$

To lowest order, the renormalization condition for the T-matrix at scale μ gives us

$$\langle \mathbf{k}'|\hat{T}(\mu)|\mathbf{k}\rangle = g_{1,R}\mathbf{k}' \cdot \mathbf{k}, \tag{3.89}$$

which, on the energy shell, implies

$$g_{1,R} = \frac{12\pi\hbar^2 a_1}{m}, \tag{3.90}$$

that is, $g_{1,R}$ is finite. In equation (3.90), a_1 is the p-wave volume. To second order, the renormalized T-matrix takes the form

$$\langle \mathbf{k}'\left|\hat{T}(z)\right|\mathbf{k}\rangle = \left[g_{1,R} + g_{3,R}\left(k'^2 + k^2\right) + A_p + B_p\left(k'^2 + k^2\right) + C_p k'^2 k^2\right]\mathbf{k}' \cdot \mathbf{k}. \tag{3.91}$$

Above, A_p, B_p and C_p are constants, while $g_{3,R}$ is a renormalized coupling constant to be determined. The fact that the T-matrix in equation (3.91) is proportional to $\mathbf{k}' \cdot \mathbf{k}$ is proven by noticing that

$$\int \frac{d^3q}{(2\pi)^3} F(q)(\mathbf{k}' \cdot \mathbf{q})(\mathbf{q} \cdot \mathbf{k}) = \int \frac{d^3q}{(2\pi)^3} F(q) \sum_{i=x,y,z} k_i' k_i q_i^2. \tag{3.92}$$

Since all integrals (with $i = x, y, z$) above are identical, we have

$$\int \frac{d^3q}{(2\pi)^3} F(q)(\mathbf{k}' \cdot \mathbf{q})(\mathbf{q} \cdot \mathbf{k}) = \mathbf{k}' \cdot \mathbf{k} \int \frac{d^3q}{(2\pi)^3} F(q) \frac{q^2}{3}. \tag{3.93}$$

The three constants A_p, B_p and C_p in equation (3.91) are given by

$$A_p = \frac{1}{3} \int \frac{d^3q q^2}{(2\pi)^3} G_R(z, \mu; \mathbf{q}) \left[g_{1,R}^2 + 2g_{1,R} g_{3,R} q^2 + g_{3,R}^2 q^4 \right], \tag{3.94}$$

$$B_p = \frac{1}{3} \int \frac{d^3q q^2}{(2\pi)^3} G_R(z, \mu; \mathbf{q}) \left[g_{1,R} g_{3,R} + g_{3,R}^2 q^2 \right], \tag{3.95}$$

$$C_p = \int \frac{d^3q q^2}{(2\pi)^3} G_R(z, \mu; \mathbf{q}) g_{3,R}^2. \tag{3.96}$$

As with the s-wave problem to NLO, we only have coupling up to $(k'^2 + k^2)$, and therefore C_p must vanish. This requires $g_{3,R}^2 \Lambda \to 0$ as $\Lambda \to \infty$. Finiteness of the constant B_p requires a finite limit $\lim_{\Lambda \to \infty} g_{3,R}^2 \Lambda^3$, that is, $g_{3,R}^2$ must vanish as $\Lambda^{-3/2}$ or faster. For A_p to be finite (and non-zero) *for all* energies z, $g_{3,R}$ has to vanish as Λ^{-2} for large cutoff values, which implies that both B_p and C_p vanish, in analogy to what we saw in the s-wave channel. Defining now $g_{3,R} = \alpha_{3,R} \Lambda^{-2}$, we see that $\alpha_{3,R}$ must satisfy

$$\alpha_{3,R}^2 + 2g_{1R}\alpha_{3R} \frac{I_{4,R}(z)}{I_{6,R}(z)} + g_{1,R}^2 \frac{I_{2,R}(z)}{I_{6,R}(z)} = 0. \tag{3.97}$$

The above relation must hold *for all z*, for otherwise A_p would be divergent for almost all z. Fortunately, the renormalized integrals are all actually z-independent (and finite) in the limit $\Lambda \to \infty$, which is due to the fact that, without the cutoff power factor in front of them, their leading-order divergence is z-independent. We slightly relax equation (3.97) to

$$\alpha_{3,R}^2 + 2g_{1R}\alpha_{3R} \frac{I_{4,R}}{I_{6,R}} + g_{1,R}^2 \frac{I_{2,R}}{I_{6,R}} = \frac{\lambda_R}{\Lambda}, \tag{3.98}$$

with λ_R a finite constant to be determined, as $\Lambda \to \infty$, which is needed to obtain full, independent tuneability of the p-wave volume and p-wave range. Since $B_p = C_p = 0$ in equations (3.95) and (3.96), then what we need to determine λ_R is the low-energy expansion of $A_p(z)$. This is fairly easy, since to $O(z)$, we have, using cutoff regularization,

$$A_p \approx -\frac{(m/\hbar^2)^2}{3}\frac{\lambda_R}{2\pi^2}z. \tag{3.99}$$

.Using now $z = \hbar^2 k^2/m$, and expanding the low-energy T-matrix, we obtain

$$\left(\frac{m}{6\pi\hbar^2}\right)^2 \lambda_R = -\pi a_1^2 r_1, \tag{3.100}$$

which relates the coupling constant $g_{3,R}$ to the physical scales a_1 and r_1.

3.3 Non-perturbative renormalization

Until now, we have studied, in detail, how to perform renormalized perturbation theory, or to renormalize ill-defined perturbative expansions in scattering theory. Since non-relativistic quantum mechanics is relatively simple, we can actually try and renormalize most low-energy theories non-perturbatively, i.e., exactly. When doing so, we will see that in fact, we had well defined theories all along, and we were simply rearranging series that should have been written in their renormalized form to begin with. This does not mean, however, that perturbative renormalization is not useful. In fact the opposite is true: non-perturbative renormalization gives a theoretical background as to *why* we are allowed to perturbatively renormalize these low-energy theories.

3.3.1 What renormalization is *really* about

Let us begin with the operator equation (3.63). Its solution is, formally

$$\hat{O}(z) = [1 - \hat{K}(z)]^{-1}\hat{W}_0. \tag{3.101}$$

The renormalization condition we have used time and again in perturbation theory would read here

$$\hat{O}(\mu) = \hat{W}_R, \tag{3.102}$$

which immediately relates the *bare* coupling operator $\hat{W}_0$ with the renormalized one as

$$\hat{W}_0 = [1 - \hat{K}(\mu)]\hat{W}_R. \tag{3.103}$$

The wise reader will note that equations (3.102) and (3.103) contain no new information as compared with equation (3.101), from a non-perturbative perspective. In fact, in order for the renormalization condition (3.102) to make any sense, equation (3.101) must be well defined. That is, the original problem that we aim to solve must be well defined in order for perturbative renormalization to make any sense at all. A valid question, therefore, is whether renormalization is useful at all non-perturbatively or not?

The question above can be answered, at least in some simple examples, positively. Let us take a well defined scattering problem determined by a quantum Hamiltonian H. The asymptotic trajectories are those associated with a part H_0 of H, and the interaction is defined as $V = H - H_0$. If H is a single-particle Hamiltonian or a two-body relative Hamiltonian, then the Lippmann–Schwinger equation is of the same kind as equation (3.101), with

$$\hat{O}(z) = \hat{T}(z), \tag{3.104}$$

$$\hat{K}(z) = VG_0(z), \tag{3.105}$$

$$\hat{W}_0 = V. \tag{3.106}$$

Clearly, there is no need to regularize or renormalize the above problem since, by definition, it is well defined. In fact, if $\mu \to \infty$, then $\hat{W}_R = V$, which is the Born limit.

Renormalization has a different purpose in a non-perturbative context. We wish to simplify the *a priori* arbitrary interaction V, so that we may be able to solve, if approximately, more complicated problems than the two-body problem (such as the three- and many-body problems), which may be extremely complicated if the full interaction V is used. Therefore, we expand the interaction V in some convenient way, as

$$V = \sum_{j=1}^{\infty} \lambda^j V_j. \tag{3.107}$$

We truncate the expansion above up to a given order n, and we work with the truncated Hamiltonian H_n, given by

$$H_n = H_0 + \sum_{j=1}^{n} \lambda^j V_j. \tag{3.108}$$

For simplicity, and (actually) without loss of generality, we can take $n = 1$ and $\lambda = 1$, so that $H_1 = H_0 + V_1$. We now attempt to solve for the T-matrix corresponding to H_1, which we call $\hat{T}_1(z)$, in the hope that the full T-matrix, on-shell, can be approximated by $\hat{T}_1(z)$, that is, we hope that

$$\frac{|\langle \mathbf{k}'|\hat{T}_1(z)|\mathbf{k}\rangle - \langle \mathbf{k}'|\hat{T}(z)|\mathbf{k}\rangle|}{|\langle \mathbf{k}'|\hat{T}(z)|\mathbf{k}\rangle|} \ll 1, \tag{3.109}$$

for $k' = k$ and $z = \hbar^2 k^2/m + i0^+$. The problem when expanding V is that, while some relevant matrix elements of V and V_1 may be very close to one another, the expansion is an operator expansion and, therefore, modifies the interaction at *all* scales. This may introduce highly non-trivial errors in the supposedly approximated T-matrix, and even infinities, non-perturbatively. On-shell, we may then renormalize the Lippmann–Schwinger equation for H_1,

$$\hat{T}_1(z) = V_1 + V_1 G_0(z) T_1(z), \tag{3.110}$$

by enforcing equation (3.109) at and around a certain energy scale μ. To do this, we express the matrix elements of V_1 as

$$\langle \mathbf{k}'|V_1|\mathbf{k}\rangle = g_0^{(0)}F_0(\mathbf{k}', \mathbf{k}) + g_0^{(1)}F_1(\mathbf{k}', \mathbf{k}) + \cdots. \tag{3.111}$$

Formally, equation (3.111), when inserted into equation (3.110), yields a T-matrix

$$\langle \mathbf{k}'|\hat{T}_1(z)|\mathbf{k}\rangle = \tilde{\tau}(\{g_0^{(j)}\}_j; z; \mathbf{k}', \mathbf{k}), \tag{3.112}$$

which, on the energy shell, becomes

$$\langle \mathbf{k}'|\hat{T}_1(z)|\mathbf{k}\rangle \equiv \tau(\{g_0^{(j)}\}_j; k, \Omega_{\mathbf{k}',\mathbf{k}}), \tag{3.113}$$

where $\Omega_{\mathbf{k}',\mathbf{k}}$ represents the angular dependence, and the dependence on energy and modulus of $\mathbf{k}$ is encoded in just k. The renormalization condition, instead of using the full T-matrix, uses on-shell information where the T-matrix is restricted to having the functional form given by equation (3.112), but with all coupling constants $g_0^{(j)}$ replaced with $\bar{g}_0^{(j)}$, in such a way that the on-shell T-matrix at energy scale μ is reproduced exactly. Hence, instead of V_1, we shall use a new operator $\bar{V}_1$ given by

$$\langle \mathbf{k}'|\bar{V}_1|\mathbf{k}\rangle = \bar{g}_0^{(0)}F_0(\mathbf{k}', \mathbf{k}) + \bar{g}_0^{(1)}F_1(\mathbf{k}', \mathbf{k}) + \cdots. \tag{3.114}$$

If we call $\hat{T}_{1,R}(\mu)$ the T-matrix which, on-shell, satisfies $\langle \mathbf{k}'|\hat{T}_{1,R}(\mu)|\mathbf{k}\rangle = \langle \mathbf{k}'|\hat{T}(\mu)|\mathbf{k}\rangle$, and takes the functional (off-shell) form in equation (3.113), with $g_0^{(j)}$ replaced with $\bar{g}_0^{(j)}$, then

$$\bar{V}_1 = [1 + G_0(\mu)\hat{T}_{1,R}(\mu)]^{-1}\hat{T}_{1,R}(\mu), \tag{3.115}$$

which allows us to write the Lippmann–Schwinger equation for $\hat{T}_{1,R}(z)$ as

$$\hat{T}_{1,R}(z) = \hat{W}_R + \hat{W}_R G_R(z, \mu)\hat{T}_{1,R}(z), \tag{3.116}$$

where we have used the renormalization condition $\hat{T}_{1,R}(\mu) = \hat{W}_R$, and where, as in the previous section,

$$G_R(z; \mu) = G_0(z) - G_0(\mu). \tag{3.117}$$

Admittedly, we have taken an extremely long detour to end up with something that, seemingly, we could have done from the start. However, the non-trivial points here are (1) the fact that at the renormalization scale, $\hat{T}_{1,R}(\mu)$ equals, on-shell, the *actual* on-shell T-matrix at the same scale, and (2) that the off-shell structure of the renormalized *operator* $\hat{W}_R = \hat{T}_{1,R}(\mu)$ coincides, functionally, with that of the *bare* interaction V_1, which we had obtained from a naïve expansion of the original interaction V. This is very important conceptually. In practice, however, this is simpler: we regularize (if necessary) the integrals involved using a momentum cutoff, use the bare interaction $\bar{V}_1$, and fix the constants $\bar{g}_0^{(j)}$ of the theory to reproduce the on-shell T-matrix at scale μ, and around it—usually by fitting scattering phase shifts—regardless of the shape of the off-shell T-matrix. We finally check for consistency, i.e., we check that the resulting $\hat{T}_{1,R}$ is non-trivial, that is, it is finite for $z \neq \mu$.

3.3.2 Counterterms

We continue here with some rather conceptual points in both non-perturbative and perturbative renormalization. In relativistic quantum field theory (QFT), the bare Lagrangian is typically supplemented with so-called counterterms which have the purpose of renormalizing the theory. In QFT, infinities are usually intrinsic to the theories, and these have to be removed consistently by regularizing and renormalizing either perturbatively or non-perturbatively. We note that, fortunately, the renormalization procedure in QFT, at least in perturbative quantum electrodynamics (QED), is *exactly* identical to what we have developed in this chapter so far. This is easy to see by inspecting any textbook in QFT, for instance [6]. In non-relativistic, low-energy effective field theories (EFT) that we target in this chapter, the original theories are usually very well defined, and it is us who, in an attempt of simplifying the theory, introduce ill-defined operators and quantities that then have to be fixed by means of renormalization. But, in essence, it is the same problem, which can be summarized as follows

- we want to describe experimental low-energy phenomenology
- we know the approximate form and coupling constants of the interactions that reproduce the desired phenomenology
- the coupling constants in this approximate form give incorrect results
- we add or subtract parts of the *full* known or unknown interaction that do not change the functional form of the approximate interaction
- the addition or subtraction generates well defined results that reproduce known phenomenology
- the renormalized theory, as incomplete as may be, is used to describe *a priori* unknown physics beyond the renormalization scale.

In the non-relativistic case, the above can be easily formalized. For instance, if the Hamiltonian is $H = H_0 + V$ and we choose $H_1 = H_0 + V_1$ as the bare Hamiltonian, then obviously in the previous subsection we added, as a counterterm, $C = \bar{V}_1 - V_1$. Obviously, $V \neq C + V_1$, and therefore the theory, even if predictive, is approximate. Since we know the full Hamiltonian, however, we could use the counterterm $C' = V - V_1$ but, in that case, the form of the interaction would change, that is, the fourth bullet point above would not be satisfied. Note that we want this to be satisfied because we wish to simplify the interactions for whichever reason.

Now, counterterms in non-relativistic EFT are not necessarily of the potential form. Also, the splitting of the Hamiltonian needs not be of the form kinetic + potential energy. There is a particularly neat example of this, namely Luttinger liquids, where the kinetic energy is approximated linearly. Therefore, we may split the original Hamiltonian $H = H_0 + V$ such that $H_0 = H_0^{(1)} + [H_0 - H_0^{(1)}]$ and $V = V_1 + [V - V_1]$. The bare Hamiltonian is now given by

$$H_1 = H_0^{(1)} + V_1, \tag{3.118}$$

and *both* $H_0^{(1)}$ and V_1 include bare coupling constants. The simplest problem of this kind, which we use here as an example, is the scattering of one particle off a fixed impurity, using linearization in one spatial dimension of the kinetic energy around two (Fermi) points $\pm k_{F,0}$. In the momentum representation, the kinetic energy, which is quadratic in k, is approximated as

$$\begin{aligned}\frac{\hbar^2 k^2}{2m} &= \frac{\hbar^2 k_{F,0}^2}{2m} + \frac{\hbar^2 k_{F,0}}{m}(|k| - k_{F,0}) + O((k - k_{F,0})^2) \\ &\equiv \epsilon_1(k) + O((k - k_{F,0})^2).\end{aligned} \tag{3.119}$$

Note that we insist on writing $k_{F,0}$ with a subscript '0', since the operator $H_0^{(1)}$ has been expanded and, upon inclusion of interactions, $k_{F,0}$ does not necessarily represent the physical Fermi point k_F. Then, $H_0^{(1)}$ is defined by its matrix elements

$$\langle k'|H_0^{(1)}|k\rangle = 2\pi\delta(k' - k)\epsilon_1(k). \tag{3.120}$$

For this example, we can use a constant in momentum space for V_1, which is regular in the original (without the linear approximation) problem. We define it as $\langle k'|V_1|k\rangle = g_0$.

We begin by renormalizing the scattering problem non-perturbatively. To include counterterms, we change H_1 to

$$H_1(c_a, c_b) = H_0^{(1)} + V_1 + c_a H_0^{(1)} + c_b V_1 = (1 + c_a)H_0^{(1)} + (1 + c_b)V_1. \tag{3.121}$$

The T-matrix is a constant $t(z)$ in momentum space, and is given by

$$\frac{1}{t_1(z)} = \frac{1}{(1 + c_b)g_0} - \int \frac{dq}{2\pi} \frac{1}{z - (1 + c_a)\epsilon_1(q)}. \tag{3.122}$$

The natural renormalization scale is the Fermi energy $\mu = \hbar^2 k_F^2/2m + i0^+$, where k_F is the actual Fermi momentum. At that scale, the non-relativistic on-shell T-matrix takes the form

$$\frac{1}{t(\mu)} = \frac{m}{\hbar^2}\left[-\frac{1}{k_F \cot\theta_{k_F}} + i\frac{1}{k_F}\right]. \tag{3.123}$$

Above, θ_{k_F} is the even-wave phase shift at momentum k_F. In equation (3.122), the integral needs regularization, and we use a hard cutoff Λ. This gives

$$\begin{aligned}\int \frac{dq}{2\pi} \frac{1}{\mu - (1 + c_a)\epsilon_1(q)} &= -(1 + c_a)\frac{\hbar^2 k_{F,0}}{m} \ln\left|\frac{2(1 + c_a)k_{F,0}\Lambda}{k_F^2 + (1 + c_a)k_{F,0}^2}\right| \\ &\quad - i\frac{m}{\hbar^2(1 + c_a)k_{F,0}}.\end{aligned} \tag{3.124}$$

Inserting the value of the integral above into equation (3.122) at $z = \mu$ and comparing with the renormalized value, equation (3.123) gives the counterterm c_a (see the imaginary parts) as

$$(1 + c_a)k_{F,0} = k_F. \tag{3.125}$$

As promised, $k_{F,0}$ does not necessarily equal its physical value k_F. The renormalization, after defining

$$\frac{1}{g_R} = -\frac{m}{\hbar^2 k_F \cot \theta_{k_F}}, \tag{3.126}$$

reduces to

$$\frac{1}{(1 + c_b)g_0} + \frac{\hbar^2 k_F}{m} \ln \left| \frac{2\Lambda}{k_F + k_{F,0}} \right| = \frac{1}{g_R}, \tag{3.127}$$

Since $k_{F,0}$ is a bare coupling constant, and the only other unphysical scale left in the problem is the cutoff, we can choose it as $k_{F,0} = 2\Lambda$ (the factor 2 is arbitrary and changing it to any other number corresponds to a mere rescaling of g_0), or equivalently we set $c_a = k_F/2\Lambda - 1$. Then, we obtain, from equation (3.127),

$$(1 + c_b)g_0 = g_R. \tag{3.128}$$

The choice of bare Fermi momentum above is actually important. The T-matrix at $z = E + i0^+ \neq \mu$ takes the form

$$\frac{1}{t_1(z)} = \frac{1}{(1 + c_b)g_0} + \frac{\hbar^2 k_F}{m} \ln \left| \frac{\hbar^2 k_F \Lambda / m}{E + \hbar^2 k_F k_{F,0}/2m} \right| + i\frac{m}{\hbar^2 k_F}. \tag{3.129}$$

By fixing $1/(1 + c_b)g_0 \sim 1/g_R - \hbar^2 k_F \ln|\Lambda|/m$, we still have a finite T-matrix. However, it depends logarithmically with the scattering energy, as

$$\frac{1}{t_1(z)} = \frac{1}{g_R} + \frac{\hbar^2 k_F}{m} \ln \left| \frac{k_F^2 + k_F k_{F,0}}{2mE/\hbar^2 + k_F k_{F,0}} \right| + i\frac{m}{\hbar^2 k_F}. \tag{3.130}$$

Moreover, we are left with a bare parameter, $k_{F,0}$, whose value is still arbitrary. If we choose $c_a = 0$, then $k_{F,0} = k_F$, and therefore the inverse of the T-matrix diverges logarithmically both in the infrared limit ($E \to 0$) and for large energies. The superior choice that introduces no new issues is definitely $c_a = k_F/2\Lambda - 1$. With this, the final form of the T-matrix, as $\Lambda \to \infty$, is

$$t_{1,R}(z) = \frac{1}{1/g_R + im/\hbar^2 k_F}, \tag{3.131}$$

which is, actually, fully energy-independent, while the original T-matrix has energy dependence in the imaginary part of the denominator, which can change as $\sim k$, while in the linear approximation, it is fixed as $\sim k_F$ for kinematic reasons.

3.3.3 Non-perturbative renormalization of low-energy interactions

Now that all methodology and concepts in renormalization have been introduced, we are ready to renormalize low-energy interactions non-perturbatively. We will work out the renormalization of the LO+NLO interactions in the s-wave sector, which is the most relevant for our purposes, and was renormalized perturbatively in section 3.2.5.

The bare potential is given by equation (3.74), which we write once more

$$\langle \mathbf{k}'|V|\mathbf{k}\rangle = g_0 + g_2(k'^2 + k^2). \tag{3.132}$$

Above, we are assuming the potential is cut off for large momenta, i.e., equation (3.132) in fact includes a factor $\theta(\Lambda - k)\theta(\Lambda - k')$. Then, g_0 and g_2 are cutoff-dependent or, equivalently, we may introduce counterterms which are allowed to depend on Λ. Solving the Lippmann–Schwinger equation for the non-perturbative T-matrix is straightforward. The T-matrix takes the form

$$\langle \mathbf{k}'|\hat{T}(z)|\mathbf{k}\rangle = t_{0,\,0}(z) + t_{0,\,2}(z)k^2 + t_{2,\,0}(z)k'^2 + t_{2,\,2}(z)k'^2k^2, \tag{3.133}$$

where $t_{i,j}(z)$ are (energy-dependent) constants. Firstly, we need $t_{2,\,2} = 0$, just as in the perturbative case. It is totally trivial to show, as was the case perturbatively, that the renormalization procedure implies $t_{0,\,2} = t_{2,\,0} = 0$ as well, and we are left with $\langle \mathbf{k}'|\hat{T}(z)|\mathbf{k}\rangle = t_{0,\,0}(z)$. The T-matrix takes the following explicit form

$$t_{0,\,0} = \frac{g_0(1/g_2 - J_2) + g_0J_2 + g_2J_4}{(1/g_2 - J_2)(1 - g_0J_0 - g_2J_2) - (g_0J_2 + g_2J_4)J_0}, \tag{3.134}$$

where $J_n = J_n(z)$ are defined as

$$J_n(z) = \int_\Lambda \frac{d^3q}{(2\pi)^2} G_0(z;\, \mathbf{q})q^n. \tag{3.135}$$

We renormalize the problem by setting

$$\mathrm{Re}\left(\frac{1}{t_{0,\,0}(0)}\right) = \frac{1}{g_{0,R}}, \tag{3.136}$$

$$\mathrm{Re}\left(\frac{1}{t_{0,\,0}(\mu)}\right) = \frac{1}{g_R(\mu)} \tag{3.137}$$

Careful inspection of equation (3.134) tells us that the leading order cutoff dependencies of the bare coupling constants g_0 and g_2 are

$$g_0 = a_0\Lambda^{-1} + b_0\Lambda^{-2} + \cdots, \tag{3.138}$$

$$g_2 = a_2\Lambda^{-3} + b_2\Lambda^{-4} + \cdots,\,, \tag{3.139}$$

where a_0, b_0, a_2 and b_2 are finite constants. We eliminate g_0 as

$$g_0 = \frac{g_{0,R}(1 - g_2J_2(0))^2 - g_2^2J_4(0)}{1 + g_{0,R}J_0(0)(1 + g_2^2J_4(0))}, \tag{3.140}$$

For large Λ, we have

$$g_2^2 J_4(0) \approx -\frac{ma_2}{10\pi^2\hbar^2\Lambda}, \tag{3.141}$$

$$g_2 J_2(0) \approx -\frac{ma_2}{6\pi^2\hbar^2}, \tag{3.142}$$

and therefore we have

$$a_0 = -\frac{2\pi^2\hbar^2}{m}\left(1 + \frac{ma_2}{6\pi^2\hbar^2}\right)^2. \tag{3.143}$$

Therefore, the lowest-order bare interactions do not really have any physical content, and exist to make quantities finite. This is consistent with the rearrangement view on renormalization. In fact, it is simpler to avoid altogether the leading order divergences, which can be removed consistently in this way since the leading order divergences of $J_n(z)$ are all energy-independent. We may simply replace $J_n(z)$ with $J_{n,R}(z)$, where

$$J_{n,R}(z) = \int_\Lambda \frac{d^3q}{(2\pi)^3} G_R(z, 0; \mathbf{q})q^n, \tag{3.144}$$

where the renormalization scale has been chosen to be zero for simplicity. Doing so, by construction (see earlier sections), the coupling constant g_0 should be finite and equal to $g_{0,R}$. This is easily checked, in equation (3.140), since $J_n(0) = 0$ for all n, and we have, indeed, $g_0 = g_{0,R}$. Then, g_2 is replaced with $g_{2,R}$ (which should still vanish for large Λ), and the condition (3.137), together with equation (3.140), becomes

$$g_{0,R} = \frac{g_R(\mu)(1 - g_{2,R}I_{2,R}(\mu))^2 - g_{2,R}^2 I_{4,R}(\mu)}{1 + g_{0,R}I_{0,R}(\mu)(1 + g_{2,R}^2 I_{4,R}(\mu))}. \tag{3.145}$$

Above, we have defined

$$I_{n,R}(z) = \mathrm{P}\int_\Lambda \frac{d^3q}{(2\pi)^3} G_R(z, 0; \mathbf{q})q^n. \tag{3.146}$$

The full, explicit solution to the problem is of little use, except for numerical value. In problem 2, the reader is asked, instead, to show that the effective range must be negative as $\Lambda \to \infty$, just as in the perturbative expansion.

3.3.4 Renormalization group invariance

We now give a brief account of how the renormalization procedures explained throughout this chapter link to the well-known renormalization group. Here, renormalization group invariance refers to the invariance, or independence, of the T-matrix when the renormalization (sliding) scale μ is changed.

In a non-perturbative setting, the Lippmann–Schwinger equation for the T-matrix is given by equation (3.116). Renaming for convenience $\hat{T}_{1,R}(z) \equiv \hat{T}_{1,R}(z;\, \mu)$, and $\hat{W}_R \equiv \hat{W}_R(\mu)$ we must prove that

$$\frac{\partial \hat{T}_{1,R}(z;\,\mu)}{\partial \mu} = 0. \tag{3.147}$$

To do this, it suffices to prove that

$$\frac{\partial \hat{W}_R}{\partial \mu}(\mu) = \hat{W}_R(\mu)\frac{\partial G_0}{\partial \mu}(\mu + i0^+)\hat{W}_R(\mu). \tag{3.148}$$

The above is true for z-independent $\hat{W}_R$, and is shown to be sufficient for showing renormalization group invariance by using equation (3.116) and setting

$$\left.\frac{\partial \hat{T}_{1,R}(z;\,\mu)}{\partial \mu}\right|_{z=\mu+i0^+} = 0. \tag{3.149}$$

Let us prove renormalization group invariance in the simplest case, that is, a momentum independent T-matrix that only depends on energy, corresponding to the description of the s-wave scattering length a. We set $\langle \mathbf{k}'|\hat{W}_R(\mu)|\mathbf{k}\rangle \equiv g_R(\mu)$. Equation (3.149) becomes

$$g_R'(\mu) = -f(\mu)g_R^2(\mu), \tag{3.150}$$

where f is given by

$$f(\mu) = \int \frac{d\mathbf{q}}{(2\pi)^3}\left(\frac{1}{\mu - \hbar^2 q^2/m + i0^+}\right)^2. \tag{3.151}$$

The solution to the renormalization group equation, equation (3.150) is simply

$$\frac{1}{g_R(\mu)} = \frac{1}{g_R(0)} + \int_0^\mu d\bar{\mu} f(\bar{\mu}), \tag{3.152}$$

finally obtaining

$$\frac{1}{g_R(\mu)} = \frac{1}{g_R(0)} + i\frac{m}{4\pi\hbar^2}k_\mu, \tag{3.153}$$

with $\hbar^2 k_\mu^2/m = \mu$. It is straightforward to check that the 'running' of the coupling constant in equation (3.153) coincides with the direct calculation of the T-matrix, see equation (3.71) with $r = 0$.

The renormalization group equation can also be obtained perturbatively, and for more complicated renormalization procedures. For instance, in the case of renormalization via counterterms, we consider a general Hamiltonian $H_1(c_a, c_b)$ as in equation (3.121). We take the non-interacting Green's function $G_0(z;\,c_a) = [z - (1 + c_a)H_0^{(1)}]^{-1}$, and the interaction $V(c_b) = (1 + c_b)V_1$. The T-matrix satisfies the Lippmann–Schwinger equation $T(z) = V(c_b) + V(c_b)G_0(z;\,c_a)T(z)$. Now, from the theory developed in section 3.3.2, c_a and c_b depend on physical quantities (which we may call renormalized quantities) C_a and C_b, both evaluated at a certain scale μ. Hence, we can formally write $c_a = c_a(\mu)$ and $c_b = c_b(\mu)$. The renormalization group invariance requires $\partial_\mu T(z; \mu)|_{z=\mu} = 0$, just as in equation (3.147). However, since we have two coupling constants, and $T(z; \mu)$ does not depend on μ by renormalization group invariance, we must also set $\partial_\mu^2 T(z; \mu)|_{z=\mu} = 0$. In general, for n coupling constants we require $\partial_\mu^j T(z; \mu)|_{z=\mu} = 0$ ($j = 1, 2, \dots, n$).

3.4 Non-renormalisable theories

The theories we have dealt with so far are all renormalizable. That is, there are well defined procedures that allow us to remove unphysical parameters, e.g., a momentum cutoff, consistently, in such a way that the results do not depend on the particular procedure employed. In the context of low-energy interactions in non-relativistic quantum mechanics, however, renormalizability does impose some restrictions. For instance, we have seen that if we allow $\Lambda \to \infty$, then the effective range of the s-wave interaction is necessarily negative. In other words, a theory with a zero-range interaction and positive effective range is not renormalizable.

The solution typically given to theories with low-energy interactions that are non-renormalizable is to leave the cutoff scale, whether in momentum (UV cutoff) or position (short-distance cutoff) representation, fixed and finite. The theoretical predictions are then only valid well below the energy scale set by the UV cutoff. In a non-relativistic collision, this energy would be $E_\Lambda \sim \hbar^2\Lambda^2/m$. This 'truncation' at high energies can generate notable issues, however, for higher particle numbers. A particular issue is that of negative probabilities [7].

In a general context, non-renormalizable theories typically arise as incomplete low-energy descriptions of a complete, renormalizable theory. Two situations can occur when this is the case: (i) the complete theory is hard to deal with, but is known or (ii) the complete theory is unknown and we only know (so far) an incomplete low-energy theory. The first situation occurs in a non-relativistic problem for which we know, to a certain extent, a two-body interaction potential $V(\mathbf{r})$ that is well-behaved. We expand the interaction at low energies, and obtain a non-renormalizable theory. Case (ii) occurs in theories of quantum gravity, for instance. It is even possible that we may be trying to describe a system using quantum field theory, while the original theory is not necessarily a quantum field theory at all.

An example where non-renormalizability has actual physical significance will be explored in section 6.1.3, in which Klein tunnelling in graphene can be explained by means of non-renormalizability of a particular electron-impurity collision theory.

Problems

1. Carry out the third order calculation of the perturbative correction in equation (3.8).
2. Show, non-perturbatively, that the effective range in NLO non-relativistic EFT must be negative when the cutoff is removed ($\Lambda \to \infty$), see section 3.3.3.

References

[1] Lähde T and Meißner U-G 2019 *Nuclear Lattice Effective Field Theory: An Introduction* (Berlin: Springer)

[2] Aziz R A and Slaman M J 1991 An examination of *ab initio* results for the helium potential energy curve *J. Chem. Phys.* **94** 8047

[3] Delamotte B 2004 A hint of renormalization *Am. J. Phys.* **72** 170

[4] Weinberg S 1964 Perturbation theory for strong repulsive potentials *J. Math. Phys.* **5** 743

[5] Beck S, Bazak B and Barnea N 2020 Removing the Wigner bound in non-perturbative effective field theory *Phys. Lett.* **B806** 135485

[6] Zee A 2003 *Quantum Field Theory in a Nutshell* (Princeton, NJ: Princeton University Press)

[7] Nishida Y 2012 Impossibility of the Efimov effect for *p*-wave interactions *Phys. Rev.* **A86** 012710

Chapter 4

Atoms, electrons and photons in structured media

This chapter switches gears completely with respect to what has been discussed in the previous chapters. We introduce a number of quantum systems that are to be regarded as physical implementations of interacting theories and models. In particular, we focus on ultracold atoms interacting with external fields, electrons in solid state materials and photons in certain structured media. The list of topics chosen is definitely incomplete, and is meant to give examples of what one can or may be able to implement with current technology.

4.1 Ultracold atoms

In this section, we consider atoms at extremely low temperature, so that quantum effects are noticeable. We focus on atom–light interactions, which are able to produce a number of interesting physical situations.

4.1.1 Quantum optics

We begin by introducing light–atom interactions. We focus on a neutral atom in the presence of a classical electric field $\mathbf{E}$, in the infinite mass approximation for the nucleus, which sits at $\mathbf{r} = 0$. For a more detailed discussion, we refer the reader to the excellent textbook by Pethick and Smith [1]. The most relevant interaction between the atom and the electric field, for our purposes, is the coupling of the Z atomic electrons, with charge $-e$ ($e > 0$), with the field. To leading order, this is given classically by the dipole approximation $-\mathbf{d} \cdot \mathbf{E}$. Quantum mechanically, however, we need the dipole operator, which is defined as

$$\hat{\mathbf{d}} = e\sum_{i=1}^{Z}\hat{\mathbf{r}}_i, \tag{4.1}$$

doi:10.1088/978-0-7503-3087-9ch4

where $\hat{\mathbf{r}}_i$ is the position operator of electron i. If we consider spherically symmetric atoms, then the dipole moment in the ground state vanishes, that is, $\langle \hat{\mathbf{d}} \rangle = 0$. However, second-order perturbation theory shifts the ground state energy in the presence of an electric field. This is called Stark shift. To see this, consider a static electric field $\mathbf{E}(\mathbf{r})$. We call the bound levels of the atom $|n\rangle$, with eigenenergies $\mathcal{E}_n$, ignore the continuum states (ionized atom), and define $\hat{V}_d = -\mathbf{d} \cdot \mathbf{E}$. Then, the Stark shift Δ is given by

$$\Delta = -\sum_{n \neq 0} \frac{|\langle n | \hat{V}_d | 0 \rangle|^2}{\mathcal{E}_n - \mathcal{E}_0}. \tag{4.2}$$

We further define $\mathbf{E} = |\mathbf{E}|\mathbf{u}$, where $\mathbf{u}$ is a unit vector in the direction of the electric field, and the static polarizability of the atom α, as

$$\alpha \equiv 2 \sum_{n \neq 0} \frac{|\langle n | \hat{\mathbf{d}} \cdot \mathbf{u} | 0 \rangle|^2}{\mathcal{E}_n - \mathcal{E}_0}. \tag{4.3}$$

Assuming the electric field is slowly varying, the Stark shift (4.2) becomes akin an external potential for the atom, $\Delta = \Delta(\mathbf{r})$, and can then be simply expressed as

$$\Delta(\mathbf{r}) = -\frac{\alpha}{2} |\mathbf{E}(\mathbf{r})|^2. \tag{4.4}$$

We now consider a more realistic situation, where the electric field is time-dependent. If this is the electric field of a laser beam, then it is essentially monochromatic, and we call its frequency ω. The field $\mathbf{E}$ can be written as

$$\mathbf{E}(\mathbf{r}, t) = \mathbf{E}^{(+)}(\mathbf{r}) e^{-i\omega t} + \mathbf{E}^{(-)}(\mathbf{r}) e^{i\omega t}. \tag{4.5}$$

The electric field must be real, which implies the constraint $[\mathbf{E}^{(+)}]^* = \mathbf{E}^{(-)}$.

Since the field is time-dependent, we need to take a short detour into time-dependent perturbation theory. The time-dependent wave function $|\psi(t)\rangle$ can be expanded on the basis of the static (field-free) atomic levels $|n\rangle$, ignoring once more ionized states. The expansion takes the form

$$|\psi(t)\rangle = \sum_n c_n(t) e^{-i\mathcal{E}_n t/\hbar} |n\rangle, \tag{4.6}$$

where the coefficients $c_n(t)$ admit the perturbative expansion

$$c_n(t) = c_n^{(0)} + c_n^{(1)}(t) + \cdots. \tag{4.7}$$

Above, $c_n^{(0)}$ label the initial state, and the leading order correction, which is time-dependent, takes the form

$$c_n^{(1)}(t) = -\frac{i}{\hbar} \sum_m \int_0^t dt' \langle n | \hat{V}_d(t') | m \rangle c_m^{(0)} e^{-i(\mathcal{E}_m - \mathcal{E}_n)t/\hbar}. \tag{4.8}$$

Let us calculate the correction in equation (4.8), considering real $\mathbf{E}^{(+)}$ for simplicity, so that we can write $\mathbf{E}(\mathbf{r}, t) = E_0 \cos(\omega t)\mathbf{u}$. The atomic transition frequencies (from the ground state) will be denoted by ω_n (>0), defined via $\hbar\omega_n = \mathcal{E}_n - \mathcal{E}_0$. We have, from equation (4.8),

$$\begin{aligned} c_n^{(1)}(t) &= \frac{iE_0}{2\hbar}\int_0^t dt'\langle n|\hat{\mathbf{d}}\cdot\mathbf{u}|0\rangle[e^{i(\omega+\omega_n)t'} + e^{-i(\omega-\omega_n)t'}] \\ &= \frac{E_0}{2\hbar}\langle n|\hat{\mathbf{d}}\cdot\mathbf{u}|0\rangle\left[\frac{e^{i(\omega+\omega_n)t}-1}{\omega+\omega_n} - \frac{e^{-i(\omega-\omega_n)t}-1}{\omega-\omega_n}\right]. \end{aligned} \tag{4.9}$$

Above, we have used as initial state the ground state of the atom, i.e., $c_m^{(0)} = \delta_{m,0}$. In equation (4.9), we clearly see that for perturbation theory to be valid, the laser frequency ω must be highly off-resonant, that is, it must be far from any atomic frequency.

We aim at describing the system statically, that is, we are interested in the time-averaged Stark shift. The period T of the electric field is simply $T = 2\pi/\omega$. Considering the ground state amplitude $a_0(t) = \exp(i\phi_0(t))$, with initial state $a_n(0) = \delta_{n,0}$, we time-average the quantity $d\phi_0/dt$, so that the energy shift is given by $\Delta E = -\hbar\langle d\phi_0/dt\rangle$, obtaining

$$\Delta E = -\frac{\alpha(\omega)}{4}E_0^2, \tag{4.10}$$

where the dynamic polarizability of the atom is given by

$$\alpha(\omega) = \sum_{n\neq 0}\frac{2\hbar\omega_n|\langle n|\hat{d}\cdot\mathbf{u}|0\rangle|^2}{(\hbar\omega_n)^2 - (\hbar\omega)^2}.$$

Above, we have omitted higher order ($\sim 1/(\omega_n \pm \omega)^2$) corrections.

We now apply this knowledge to a simplified model for an atom, which is valid when the laser frequency ω is near-resonant with a particular transition frequency ω_{nm} of the atom. In that case, we may consider only two electronic levels of the atom, which we shall name, following standard conventions, 'ground' and 'excited' states (which are not necessarily the true ground and first excited states). The notation we use is $|g\rangle$ and $|e\rangle$ for ground and excited state, respectively. The atomic Hamiltonian H_0, in the absence of external fields, is given by

$$H_0 = \hbar\omega_g|g\rangle\langle g| + \hbar\omega_e|e\rangle\langle e| + \text{constant} \equiv \frac{\hbar\omega_{eg}}{2}[|e\rangle\langle e| - |g\rangle\langle g|]. \tag{4.11}$$

In the last equality above, we have shifted the irrelevant constant (times the identity $=|g\rangle\langle g| + |e\rangle\langle e|$) and removed it since it does not affect the dynamics of the system in any way, as well as defined $\omega_{eg} = \omega_e - \omega_g$. In matrix notation, the operator $|e\rangle\langle e| - |g\rangle\langle g|$ takes the form

$$|e\rangle\langle e| - |g\rangle\langle g| = \begin{pmatrix} 1 & 0 \\ 0 & -1 \end{pmatrix} = \sigma_z, \tag{4.12}$$

that is, it can be written as the spin-1/2 Pauli matrix σ_z. Since we are considering the external field as classical, this has no dynamics, i.e., the microscopic atom does not influence the macroscopic electric field, and therefore we do not need to consider its Hamiltonian. The laser–atom interaction does influence the atom, and in the dipole approximation the Hamiltonian is also easy to construct. The full interaction is just $\hat{V}_d$, and in the basis of the two relevant atomic states, we have

$$\hat{V}_d = \begin{pmatrix} v_{ee} & v_{eg} \\ v_{ge} & v_{gg} \end{pmatrix}, \tag{4.13}$$

where $v_{ij} \equiv \langle i|\hat{V}_d|j\rangle$. For spherically symmetric ground and excited states, $v_{ee} = v_{gg}$, and we can write

$$\hat{V}_d = v_{ge}\sigma_- + v_{eg}\sigma_+, \tag{4.14}$$

where $\sigma_\pm$= are the ladder Pauli operators for spin-1/2. The full Hamiltonian is really simple in this case, and the dynamics is essentially trivial.

4.1.2 Optical lattices

We consider now the time-averaged Stark shift for a spatially dependent electric field,

$$\Delta E = -\frac{1}{2}\alpha(\omega)\frac{1}{T}\int_0^T |\mathbf{E}(\mathbf{r},\, t)|^2 dt. \tag{4.15}$$

When the frequency of the laser is near-resonant with an atomic transition $g \rightarrow e$, the dynamic polarizability can be approximated by

$$\alpha(\omega) = \frac{|\langle e|\hat{d}\cdot\mathbf{u}|g\rangle|^2}{\hbar\omega_{eg} - \hbar\omega}. \tag{4.16}$$

The energy shift, ΔE, of the ground state, is then spatially dependent, and we may treat it as an external potential for the atom. For a one-dimensional optical lattice, that is, a periodic potential, all we need is a standing wave laser field, which has the form

$$E(x,\, t) = E_0[\cos(kx - \omega t) + \cos(kx + \omega t)] = 2E_0\cos(kx)\cos(\omega t). \tag{4.17}$$

Introducing equation (4.17) into equation (4.15), we obtain the following effective potential, called one-dimensional optical lattice:

$$V(x) = -\alpha(\omega)E_0^2\cos^2(kx). \tag{4.18}$$

We note that, although we have considered the simplest possible case, there currently exist a large variety of geometries that can be implemented experimentally, including quasi-periodic optical lattices, i.e., synthetic quasicrystals for neutral atoms [2].

4.1.3 Synthetic vector potentials

We discuss now the possibility of generating effective vector potentials in otherwise non-magnetic systems. We begin by considering the simplest scenario of a two-level atom with ground and excited states $|g\rangle$ and $|e\rangle$, respectively. We couple the atom to a laser beam, whose interaction is parametrized as follows

$$\hat{U} = \frac{\hbar\Omega}{2}\begin{pmatrix} \cos\theta & e^{-i\phi}\sin\theta \\ e^{i\phi}\sin\theta & -\cos\theta \end{pmatrix}. \tag{4.19}$$

Above, the so-called Rabi frequency Ω is constant in space, while $\theta = \theta(\mathbf{r})$ and $\phi = \phi(\mathbf{r})$ are (importantly) space-dependent. The essence of the adiabatic approximation consists of establishing first the dressed-state basis. This is given by the space-dependent eigenstates of the laser–atom interaction at a fixed position $\mathbf{r}$. The two eigenvalues are $E_{\hat{U}}^{\pm} = \pm\hbar\Omega/2$, and the eigenstates $|\chi_{\pm}\rangle$ can be written as

$$|\chi_{+}\rangle = \begin{pmatrix} \cos(\theta/2) \\ e^{i\phi}\sin(\theta/2) \end{pmatrix}, \quad |\chi_{-}\rangle = \begin{pmatrix} -\sin(\theta/2) \\ e^{i\phi}\cos(\theta/2) \end{pmatrix}. \tag{4.20}$$

We now allow the atom to move non-relativistically, while in the presence of the laser beam. We can write the Hamiltonian H as

$$H = \frac{\mathbf{p}^2}{2m} + \hat{U}, \tag{4.21}$$

The Schrödinger equation, $H|\psi\rangle = i\hbar\partial_t|\psi\rangle$, is written in the basis of dressed states $|\chi_{\pm}\rangle$, so that

$$|\psi(\mathbf{r}, t)\rangle = \sum_{j=\pm}\psi_j(\mathbf{r}, t)|\chi_j(\mathbf{r})\rangle. \tag{4.22}$$

Applying the Hamiltonian on the wave function (4.22), projecting onto $|\chi_{+}\rangle$ and using the adiabatic approximation ($\psi_2 = 0$), leads to the equation of motion

$$i\hbar\partial_t\psi_{+} \approx \left[\frac{(\mathbf{p} - \mathbf{a})^2}{2m} + \frac{\hbar\Omega}{2} + V\right]\psi_{+}, \tag{4.23}$$

where

$$\mathbf{a} = i\hbar\langle\chi_{+}|\nabla\chi_{+}\rangle, \tag{4.24}$$

$$V = \frac{\hbar^2}{2m}|\langle\chi_{-}|\nabla\chi_{+}\rangle|^2. \tag{4.25}$$

More interesting cases than the above can be thought of. In particular, degenerate states can lead to non-Abelian (non-commutative) vector potentials in the adiabatic approximation. For example, spin–orbit coupling, in the sense of Rashba or Dresselhaus and not the 'usual' atomic, relativistic case of $\mathbf{L}\cdot\mathbf{S}$ coupling, can be generated in this way. For instance, Rashba coupling is defined by the following single-particle Hamiltonian in two spatial dimensions,

$$H = \frac{(\mathbf{p} - \lambda\sigma)^2}{2m}, \tag{4.26}$$

with $\sigma = (\sigma_x, \sigma_y)$ a vector of Pauli matrices, and λ a constant. Since $[\sigma_x, \sigma_y] \neq 0$, the vector potential $\lambda\sigma$ is non-Abelian. Since σ_i is idempotent and commutes with $\mathbf{p}_i$, the Hamiltonian (4.26) can be rewritten as

$$H = \frac{\mathbf{p}^2}{2m} - \beta\sigma \cdot \mathbf{p} + \frac{\lambda^2}{2m}, \tag{4.27}$$

where we have defined $\beta = \lambda/m$. The operator $\mathcal{H} = \sigma \cdot \mathbf{p}$ defines the helicity, that is, the momentum in the direction of the spin. For a single spin-1/2 fermion, we can diagonalize H in the basis of eigenstates of helicity. Hence, we solve

$$\sigma \cdot \mathbf{p}|\psi_{\mathbf{k}}(\mathbf{r})\rangle = h(\mathbf{k})|\psi_{\mathbf{k}}(\mathbf{r})\rangle. \tag{4.28}$$

The spinor wave function $|\psi_{\mathbf{k}}(\mathbf{r})\rangle = (\psi_{\uparrow,\mathbf{k}}, \psi_{\downarrow,\mathbf{k}})^{\mathrm{T}}$ can be written as

$$|\psi_{\mathbf{k}}(\mathbf{r})\rangle = \exp(i\mathbf{k} \cdot \mathbf{r})(\alpha_\uparrow, \alpha_\downarrow)^{\mathrm{T}}, \tag{4.29}$$

with α_i space independent. Writing $\mathbf{k}$ in polar coordinates, $\mathbf{k} = |\mathbf{k}|(\cos\phi, \sin\phi)$, equation (4.28) becomes

$$\hbar|\mathbf{k}|\begin{pmatrix} 0 & e^{-i\phi} \\ e^{i\phi} & 0 \end{pmatrix}\begin{pmatrix} \alpha_\uparrow \\ \alpha_\downarrow \end{pmatrix} = h(\mathbf{k})\begin{pmatrix} \alpha_\uparrow \\ \alpha_\downarrow \end{pmatrix}. \tag{4.30}$$

Solving equation (4.30), we obtain the two eigenvalues of the helicity at momentum $\mathbf{k}$, $h(\mathbf{k}) = \pm\hbar|\mathbf{k}|$, with eigenstates (4.29), and

$$\begin{pmatrix} \alpha_\uparrow^\pm \\ \alpha_\downarrow^\pm \end{pmatrix} = \frac{1}{\sqrt{2}}\begin{pmatrix} 1 \\ \pm\, e^{i\phi} \end{pmatrix}. \tag{4.31}$$

Since, moreover, $[H, \mathcal{H}] = 0$, we obtain two (gapless) branches for the energy

$$E_\pm(\mathbf{k}) = \frac{\hbar^2 k^2}{2m} \mp \frac{\hbar\lambda}{m}|\mathbf{k}| + \frac{\lambda^2}{2m} = \frac{1}{2m}(\hbar|\mathbf{k}| \mp \lambda)^2. \tag{4.32}$$

Many other instances of non-Abelian vector potentials can be realized taking advantage of degenerate atomic levels [3]

4.1.4 Quantization of the free electromagnetic field

To be able to study quantum effects of light interacting with atoms, we need to quantize light. That is, we need to quantize the electromagnetic field. We shall begin with the classical field theory of electromagnetism in vacuum, and then proceed to quantize it. Note that, in this section, we set the speed of light in vacuum to unity ($c = 1$).

4.1.4.1 Classical electromagnetism

We start from the Lagrangian

$$\mathcal{L} = -\frac{1}{4}F_{\mu\nu}F^{\mu\nu}, \tag{4.33}$$

where $F_{\mu\nu} = \partial_\mu A_\nu - \partial_\nu A_\mu$ is the field strength, with $A_\mu = (A_0, -\mathbf{A})$ the vector potential in relativistic notation. Above, and in the subsequent discussions, we follow Einstein's summation convention. That is, given a 4-vector $a_\mu = (a_0, \mathbf{a})$, and Minkowski's metric $\eta_{\mu\nu} = \eta^{\mu\nu}$,

$$\eta_{\mu\nu} = \begin{pmatrix} 1 & & & \\ & -1 & & \\ & & -1 & \\ & & & -1 \end{pmatrix}, \tag{4.34}$$

we have $a^\mu = \eta^{\mu\alpha}a_\alpha$, and $a_\mu b^\mu = \sum_{\mu=0}^{3}\eta_{\mu\mu}a_\mu a_\mu = a_0 b_0 - \mathbf{a}\cdot\mathbf{b}$. Developing the Lagrangian in equation (4.33) in terms of A_μ, we have

$$\mathcal{L} = -\frac{1}{2}\left(\partial_\mu A_\nu \partial^\mu A^\nu - \partial_\mu A_\nu \partial^\nu A^\mu\right). \tag{4.35}$$

The equations of motion follow from Euler–Lagrange equations

$$\frac{\partial \mathcal{L}}{\partial A_\nu} - \partial_\mu\left[\frac{\partial \mathcal{L}}{\partial(\partial_\mu A_\nu)}\right] = -\partial_\mu\left[\frac{\partial \mathcal{L}}{\partial(\partial_\mu A_\nu)}\right] = 0, \tag{4.36}$$

where in the last equality we have used the fact that $\mathcal{L}$ only depends on A_ν via $\partial_\mu A_\nu$. The derivatives we need are straightforward, and the equation of motion (4.36) becomes

$$\partial_\mu F^{\mu\nu} = 0. \tag{4.37}$$

The above equations are nothing but Maxwell's equations, without sources, in compact form. For instance, let us take $F^{0\nu}$, which is simply $F^{0\nu} = \partial^0 A^\nu - \partial^\nu A^0$, and see what it corresponds to, going step-by-step, for pedagogical value, since most of this book deals with non-relativistic formalism. We write $F^{0\nu} = (F^{00}, \mathbf{F}_c) = (0, \mathbf{F}^0)$. The three-vector $\mathbf{F}^{(0)}$ is obtained from $\mathbf{F}_0$ ($F_{0\nu} = (0, \mathbf{F}_0)$) by raising indices, so that $\partial^0 = \partial_0 = \partial/\partial t$ and $\partial^\nu = -\partial_\nu$ ($\nu = 1, 2, 3$), while $A^\mu = (A_0, \mathbf{A})$. We obtain

$$\mathbf{F}^0 = \frac{\partial \mathbf{A}}{\partial t} + \nabla A_0 = -\mathbf{E} = (E^x, E^y, E^z). \tag{4.38}$$

The corresponding Maxwell equation for $F^{\mu 0} = -F^{0\mu}$, equation (4.37), takes the form

$$0 = \partial_\mu F^{0\mu} = -\nabla \cdot \mathbf{E}, \tag{4.39}$$

which is the first of Maxwell's equations. It is easy to see, in the same way, that the field strength tensor $F_{\mu\nu}$ is given by

$$F_{\mu\nu} = \begin{pmatrix} 0 & E^x & E^y & E^z \\ -E^x & 0 & -B^z & B^y \\ -E^y & B^z & 0 & -B^x \\ -E^z & -B^y & B^x & 0 \end{pmatrix}. \tag{4.40}$$

The equations of motion also give, with the field strength tensor in equation (4.40), the Maxwell equation $\partial_t \mathbf{E} = \nabla \times \mathbf{B}$.

To obtain the remaining two Maxwell equations $\partial_t \mathbf{B} = -\nabla \times \mathbf{E}$ and $\nabla \cdot \mathbf{B} = 0$, we need to first prove that

$$\epsilon^{\alpha\beta\gamma\delta}\partial_\gamma F_{\alpha\beta} = 0(\delta = 0, 1, 2, 3), \tag{4.41}$$

where $\epsilon^{\alpha\beta\gamma\delta}$ is the four-dimensional Levi–Civita tensor. To prove equation (4.41), we write the following trivial expressions

$$\partial_\gamma F_{\alpha\beta} = \partial_\gamma \partial_\alpha A_\beta - \partial_\gamma \partial_\beta A_\alpha, \tag{4.42}$$

$$\partial_\alpha F_{\gamma\beta} = \partial_\alpha \partial_\gamma A_\beta - \partial_\alpha \partial_\beta A_\gamma, \tag{4.43}$$

$$\partial_\beta F_{\alpha\gamma} = \partial_\alpha \partial_\beta A_\gamma - \partial_\beta \partial_\gamma A_\alpha. \tag{4.44}$$

Combining equations (4.43) and (4.44), we obtain

$$\partial_\alpha F_{\gamma\beta} + \partial_\beta F_{\alpha\gamma} = \partial_\gamma F_{\alpha\beta}. \tag{4.45}$$

Because the indices on the left hand side of equation (4.41) are dummy, we may also write

$$\epsilon^{\alpha\beta\gamma\delta}\partial_\gamma F_{\alpha\beta} = \epsilon^{\gamma\beta\alpha\delta}\partial_\alpha F_{\gamma\beta} = \epsilon^{\alpha\gamma\beta\delta}\partial_\beta F_{\alpha\gamma}. \tag{4.46}$$

Introducing equation (4.45) into the left hand side of equation (4.41), and using the antisymmetry of the Levi–Civita tensor, we have

$$\epsilon^{\alpha\beta\gamma\delta}\partial_\gamma F_{\alpha\beta} = -\epsilon^{\gamma\beta\alpha\delta}\partial_\alpha F_{\gamma\beta} - \epsilon^{\alpha\gamma\beta\delta}\partial_\beta F_{\alpha\gamma}, \tag{4.47}$$

and using equation (4.46), we have equation (4.41), as we wanted to show.

Now that we have the relations (4.41), let us see what we obtain by setting $\delta = 0$. Firstly, note that equation (4.41) for $\delta = 0$ is equivalent to

$$\epsilon^{\alpha\beta\gamma}\partial_\gamma F_{\alpha\beta} = 0, \tag{4.48}$$

where all the indices are now restricted to 1, 2, 3. Expanding the sum in equation (4.48), we have

$$\begin{aligned} &\partial_3 F_{12} - \partial_3 F_{21} - \partial_2 F_{13} + \partial_2 F_{31} + \partial_1 F_{23} - \partial_1 F_{32} = 0 \\ &\Rightarrow \partial_1 F_{23} + \partial_2 F_{31} + \partial_3 F_{12} = 0. \end{aligned} \tag{4.49}$$

The above equation gives $\nabla \cdot \mathbf{B} = 0$ (see equation (4.40)), one of the two Maxwell equations remaining. It is left as a problem to show $\partial_t \mathbf{B} = -\nabla \times \mathbf{E}$. To summarize, Maxwell equations in covariant form are given by

$$\partial_\mu F^{\mu\nu} = 0, \tag{4.50}$$

$$\epsilon^{\alpha\beta\gamma\delta} \partial_\gamma F_{\alpha\beta} = 0, \tag{4.51}$$

and in standard form

$$\nabla \cdot \mathbf{E} = 0, \tag{4.52}$$

$$\frac{\partial \mathbf{E}}{\partial t} - \nabla \times \mathbf{B} = 0, \tag{4.53}$$

$$\frac{\partial \mathbf{B}}{\partial t} + \nabla \times \mathbf{E} = 0, \tag{4.54}$$

$$\nabla \cdot \mathbf{B} = 0. \tag{4.55}$$

We now take a look at gauge invariance. If we add a function $f(x)$ (x is the spacetime 4-vector) to A_μ, and define the gauge transformation

$$A'_\mu = A_\mu(x) + \partial_\mu f(x), \tag{4.56}$$

the field strength tensor $F_{\mu\nu}$ transforms to $F'_{\mu\nu}$, and

$$F'_{\mu\nu} = \partial_\mu A'_\nu - \partial_\nu A'_\mu = \partial_\mu A_\nu - \partial_\nu A_\mu + \partial_\mu \partial_\nu f - \partial_\nu \partial_\mu f = F_{\mu\nu}. \tag{4.57}$$

That is, $F_{\mu\nu}$ is gauge-invariant. Therefore, the Lagrangian $\mathcal{L}$ of electrodynamics in equation (4.33) is gauge-invariant too. This means that if A_μ and A'_μ differ in a total derivative, the dynamics of the system is unchanged: our description of the system is redundant. Another redundancy comes from the fact that A_0, that is, the electrostatic potential, is actually static, since $\partial_t A_0$ does not appear in the Lagrangian. This means that it is heavily constrained, and can be solved for by means of the first of Maxwell equations, equation (4.52). Introducing the definition, equation (4.38), of the electric field into equation (4.52), we obtain

$$0 = \nabla \cdot \left[\frac{\partial \mathbf{A}}{\partial t} + \nabla A_0 \right] = \nabla \cdot \frac{\partial \mathbf{A}}{\partial t} + \nabla^2 A_0. \tag{4.58}$$

To solve the equation above, we can solve the generic equation $\nabla^2 A_0 = f$, with f a function. This is just Poisson's equation, which we solve by introducing the Green's function $\mathcal{G}(\mathbf{x} - \mathbf{y})$, satisfying

$$\nabla^2 \mathcal{G}(\mathbf{x} - \mathbf{y}) = \delta(\mathbf{x} - \mathbf{y}). \tag{4.59}$$

Above, $\nabla = (\partial_{x_1}, \partial_{x_2}, \partial_{x_3})$ only refers to $\mathbf{x}$. Therefore, we have

$$\int d\mathbf{y}\nabla^2\mathcal{G}(\mathbf{x} - \mathbf{y})f(\mathbf{y}) = f(\mathbf{x}). \tag{4.60}$$

The above relation immediately implies

$$A_0(x) = -\int d\mathbf{y}\mathcal{G}(\mathbf{x} - \mathbf{y})\nabla_{\mathbf{y}} \cdot \frac{\partial \mathbf{A}}{\partial t}(y), \tag{4.61}$$

where A_0 and $\mathbf{A}$ are evaluated at equal times t. Note that the Green's function $\mathcal{G}$ has already been calculated in arbitrary space dimensions, as it is essentially the same as the non-interacting Green's function at zero energy that we obtained in non-relativistic scattering.

Clearly, gauge invariance and the static nature of the electrostatic potential imply that even though A_μ has four components, there are only two degrees of freedom. We will see shortly what these correspond to. We now get rid of these redundancies by picking a particular gauge, and we shall choose the so-called Coulomb gauge, in which

$$\nabla \cdot \mathbf{A} = 0. \tag{4.62}$$

Inserting the Coulomb gauge condition above, equation (4.61) implies $A_0 = 0$. This is quite convenient since it eliminates the electrostatic potential altogether.

We can finally proceed with the quantization of the electromagnetic field. We shall attempt canonical quantization now. The relevant fields in the Lagrangian are A_μ. The conjugate momenta Π^μ are given by

$$\Pi^\mu = \frac{\partial \mathcal{L}}{\partial(\partial_0 A_\mu)} = (0, \mathbf{E}), \tag{4.63}$$

that is, the momentum conjugate to A_0 vanishes, while the momentum conjugate to the vector potential $\mathbf{A}$ is the electric field $\mathbf{E}$. The Hamiltonian H of the electromagnetic field is computed as the Legendre transform

$$H = \int d\mathbf{r}\left(\Pi^\mu \partial_0 A_\mu - \mathcal{L}\right). \tag{4.64}$$

From equation (4.40) for the field strength tensor, the Lagrangian reads $\mathcal{L} = (|\mathbf{E}|^2 - |\mathbf{B}|^2)/2$. Using the Coulomb gauge condition, $\Pi^\mu \partial_0 A_\mu = |\mathbf{E}|^2$ and, therefore,

$$H = \int d\mathbf{r}\frac{1}{2}(|\mathbf{E}|^2 + |\mathbf{B}|^2). \tag{4.65}$$

Recall that in the Coulomb gauge $\mathbf{E} = -\partial_t \mathbf{A}$, and $\mathbf{B} = \nabla \times \mathbf{A}$. Hence, differentiating with respect to time and using equation (4.53), we have $\partial_t \mathbf{E} = -\partial_t^2 \mathbf{A} = \nabla \times \mathbf{B} = \nabla \times \nabla \times \mathbf{A}$. Using the identity

$$\nabla^2 \mathbf{A} = \nabla(\nabla \cdot \mathbf{A}) - \nabla \times \nabla \times \mathbf{A}, \tag{4.66}$$

and the gauge condition (4.62), we obtain the wave equation

$$\left(\nabla^2 - \frac{\partial^2}{\partial t^2}\right)\mathbf{A} \equiv \Box\mathbf{A} = -\partial_\mu\partial^\mu\mathbf{A} = 0, \tag{4.67}$$

where $\Box$ is the D'Alembertian. The wave equation (4.67) can be solved as an expansion in plane waves

$$\mathbf{A}(x) = \int \frac{d\mathbf{k}}{(2\pi)^3}\mathbf{g}(\mathbf{k})e^{i\mathbf{k}\cdot\mathbf{r}-i|\mathbf{k}|t} + \text{c. c}.. \tag{4.68}$$

The gauge condition (4.62) takes the form, in momentum representation

$$\mathbf{k}\cdot\mathbf{g}(\mathbf{k}) = 0. \tag{4.69}$$

Since $\mathbf{g}\perp\mathbf{k}$, we may choose two mutually perpendicular vectors $\mathbf{e}_{\mathbf{k},\lambda}$ ($\lambda = 1, 2$) that are perpendicular to $\mathbf{k}$, which represent the polarization states of the electromagnetic field. The polarization vectors satisfy the following completeness relation

$$\sum_{\lambda=1,2} e^i_{\mathbf{k},\,\lambda}e^j_{\mathbf{k},\,\lambda} = \delta_{ij} - \frac{k^ik^j}{\mathbf{k}^2}. \tag{4.70}$$

We prove this by noting that, since $\sum_i e^i_{\mathbf{k},\,\lambda}e^i_{\mathbf{k},\,\lambda} = 1$, then

$$\sum_\lambda\sum_i e^i_{\mathbf{k},\,\lambda}e^i_{\mathbf{k},\,\lambda} = 2, \tag{4.71}$$

and that, since $\sum_i e^i_{\mathbf{k},\,\lambda}k^i = 0$, then

$$\sum_\lambda\sum_{ij} e^i_{\mathbf{k},\,\lambda}e^j_{\mathbf{k},\,\lambda}\frac{k^ik^j}{\mathbf{k}^2} = 0. \tag{4.72}$$

Inserting the completeness relation (4.70) into the left hand sides of equations (4.71) and (4.72), we see that these are exactly satisfied.

We specify now the eigenmode decomposition (4.70) by writing

$$\mathbf{g}(\mathbf{k}) = \sum_{\lambda=1,2}\mathbf{e}_{\mathbf{k},\lambda}A_{\mathbf{k},\lambda}, \tag{4.73}$$

and defining $A_\lambda(\mathbf{k}, t) = A_{\mathbf{k},\lambda}\exp(-i|\mathbf{k}|t)$. We finally have for the field expansion

$$\mathbf{A}(x) = \int \frac{d\mathbf{k}}{(2\pi)^3}\sum_{\lambda=1,2}\mathbf{e}_{\mathbf{k},\lambda}[A_\lambda(\mathbf{k}, t)e^{i\mathbf{k}\cdot\mathbf{r}} + A^*_\lambda(\mathbf{k}, t)e^{-i\mathbf{k}\cdot\mathbf{r}}]. \tag{4.74}$$

The electric and magnetic fields read

$$\mathbf{E}(x) = -\int \frac{d\mathbf{k}}{(2\pi)^3}\sum_{\lambda=1,2}\mathbf{e}_{\mathbf{k},\lambda}[\dot{A}_\lambda(\mathbf{k}, t)e^{i\mathbf{k}\cdot\mathbf{r}} + \dot{A}^*_\lambda(\mathbf{k}, t)e^{-i\mathbf{k}\cdot\mathbf{r}}], \tag{4.75}$$

$$\mathbf{B}(x) = \int \frac{d\mathbf{k}}{(2\pi)^3} \sum_{\lambda=1,2} i\mathbf{k} \times \mathbf{e}_{\mathbf{k},\lambda}[A_\lambda(\mathbf{k}, t)e^{i\mathbf{k}\cdot\mathbf{r}} - A_\lambda^*(\mathbf{k}, t)e^{-i\mathbf{k}\cdot\mathbf{r}}]. \tag{4.76}$$

Inserting the two expressions (4.75) and (4.76) into the Hamiltonian (4.65), we obtain the energy of the electromagnetic field in terms of $A_\lambda(\mathbf{k}, t)$. We do this explicitly for the magnetic part of the energy, which is slightly more complicated, and leave the electric part as an exercise to the reader. We focus on the integral

$$\int d\mathbf{r}|\mathbf{B}|^2 = \int d\mathbf{r}\mathbf{B}^*\cdot\mathbf{B}. \tag{4.77}$$

Once expanded, equation (4.77) is simplified as follows. Firstly, note that, as a convention, we may choose $\mathbf{e}_{\mathbf{k},\lambda} = \mathbf{e}_{-\mathbf{k},\lambda}$, a choice that must be kept for the electric field as well for consistency. Then, since $\mathbf{e}_{\mathbf{k},\lambda} \cdot \mathbf{e}_{\mathbf{k},\lambda'} = \delta_{\lambda,\lambda'}$, the product of the two vectors $(\mathbf{k} \times \mathbf{e}_{\mathbf{k},\lambda}) \cdot (\mathbf{k} \times \mathbf{e}_{\mathbf{k},\lambda'}) = |\mathbf{k} \times \mathbf{e}_{\mathbf{k},\lambda}|^2\delta_{\lambda,\lambda'}$. And since $\mathbf{k}$ is perpendicular to $\mathbf{e}_{\mathbf{k},\lambda}$, we have $|\mathbf{k} \times \mathbf{e}_{\mathbf{k},\lambda}| = |\mathbf{k}|$. Putting all this together, we obtain

$$\frac{1}{2}\int d\mathbf{r}|\mathbf{B}|^2 = \sum_{\lambda=1,2} \int \frac{d\mathbf{k}}{(2\pi)^3}\mathbf{k}^2[|A_\lambda(\mathbf{k}, t)|^2 + 2\,\mathrm{Re}\,(A_\lambda(\mathbf{k}, t)A_\lambda(-\mathbf{k}, t))]. \tag{4.78}$$

The electric energy, in turn, is given by

$$\frac{1}{2}\int d\mathbf{r}|\mathbf{E}|^2 = \sum_{\lambda=1,2} \int \frac{d\mathbf{k}}{(2\pi)^3}[|\dot{A}_\lambda(\mathbf{k}, t)|^2 + 2\,\mathrm{Re}\,(\dot{A}_\lambda(\mathbf{k}, t)\dot{A}_\lambda(-\mathbf{k}, t))]. \tag{4.79}$$

From the equations of motion for $A_\lambda(\mathbf{k}, t)$, we have $A_\lambda(\mathbf{k}, t) = A_{\mathbf{k},\lambda}\exp(-i|\mathbf{k}|t)$, and therefore the Hamiltonian (4.65) simplifies to

$$H = 2\sum_{\lambda=1,2} \int \frac{d\mathbf{k}}{(2\pi)^3}\mathbf{k}^2|A_{\mathbf{k},\lambda}|^2 = 2\sum_{\lambda=1,2} \int \frac{d\mathbf{k}}{(2\pi)^3}\mathbf{k}^2|A_\lambda(\mathbf{k}, t)|^2. \tag{4.80}$$

4.1.4.2 Quantized electromagnetic field

We have developed all the necessary classical field theory of the electromagnetic field to quantize it canonically. For this, we impose the canonical commutation relations on the fields A_j and their canonically conjugate momenta, equation (4.66), $\Pi^j = \mathbf{E}^j$ ($j = 1, 2, 3$). We first rewrite the classical Hamiltonian (4.80) without assuming the commutation between $A_{\mathbf{k},\lambda}$ and its adjoint,

$$H = \sum_{\lambda=1,2} \int \frac{d\mathbf{k}}{(2\pi)^3}\mathbf{k}^2\left[A_{\mathbf{k},\lambda}A_{\mathbf{k},\lambda}^* + A_{\mathbf{k},\lambda}^*A_{\mathbf{k},\lambda}\right]. \tag{4.81}$$

We further define $\alpha_{\mathbf{k},\lambda} = (2|\mathbf{k}|)^{1/2}A_{\mathbf{k},\lambda}$, and analogously for $\alpha_{\mathbf{k},\lambda}^*$, and raise them to operators $\alpha_{\mathbf{k},\lambda} \to a_{\mathbf{k},\lambda}$, $\alpha_{\mathbf{k},\lambda}^* \to a_{\mathbf{k},\lambda}^\dagger$ with the following commutation relations

$$[a_{\mathbf{k},\lambda}, a_{\mathbf{k}',\lambda'}] = [a_{\mathbf{k},\lambda}^\dagger, a_{\mathbf{k}',\lambda'}^\dagger] = 0, \tag{4.82}$$

$$[a_{\mathbf{k},\lambda}, a^{\dagger}_{\mathbf{k}',\lambda'}] = (2\pi)^3\delta_{\lambda,\lambda'}\delta(\mathbf{k} - \mathbf{q}). \tag{4.83}$$

The quantized fields $\mathbf{A}(\mathbf{r}) \equiv \mathbf{A}(x^0 = 0, \mathbf{r})$ and their canonical momenta $\mathbf{E}(\mathbf{r}) \equiv \mathbf{E}(0, \mathbf{r})$ then become, from equations (4.74) and (4.75),

$$\mathbf{A}(\mathbf{r}) = \sum_{\lambda=1,2} \int \frac{d\mathbf{k}}{(2\pi)^3(2|\mathbf{k}|)^{1/2}} \mathbf{e}_{\mathbf{k},\lambda}\left[a_{\mathbf{k},\lambda}e^{i\mathbf{k}\cdot\mathbf{r}} + a^{\dagger}_{\mathbf{k},\lambda}e^{-i\mathbf{k}\cdot\mathbf{r}}\right], \tag{4.84}$$

$$\mathbf{E}(\mathbf{r}) = -i\sum_{\lambda=1,2} \int \frac{d\mathbf{k}}{(2\pi)^3}\left(\frac{|\mathbf{k}|}{2}\right)^{1/2} \mathbf{e}_{\mathbf{k},\lambda}\left[a_{\mathbf{k},\lambda}e^{i\mathbf{k}\cdot\mathbf{r}} - a^{\dagger}_{\mathbf{k},\lambda}e^{-i\mathbf{k}\cdot\mathbf{r}}\right]. \tag{4.85}$$

Let us compute the commutators between fields and momenta. Introducing equations (4.84) and (4.85) into the commutators $[A_i(\mathbf{r}), E^j(\mathbf{r}')] = -[A^i(\mathbf{r}), E^j(\mathbf{r}')]$, and using the commutation relations (4.82) and (4.83), we obtain

$$[A_i(\mathbf{r}), E^j(\mathbf{r}')] = -\frac{i}{2}\int \frac{d\mathbf{k}}{(2\pi)^3}\left(\sum_{\lambda} e^i_{\mathbf{k},\lambda}e^j_{\mathbf{k},\lambda}\right)[e^{i\mathbf{k}\cdot(\mathbf{r}-\mathbf{r}')} + e^{-i\mathbf{k}\cdot(\mathbf{r}-\mathbf{r}')}]. \tag{4.86}$$

We insert the completion relation for the polarization vectors, equation (4.70), and obtain

$$[A_i(\mathbf{r}), E^j(\mathbf{r}')] = -i(1 - \mathcal{G}(\mathbf{r} - \mathbf{r}')\partial^i\partial^j)\delta(\mathbf{r} - \mathbf{r}'), \tag{4.87}$$

where $\mathcal{G}$ is the Green's function of the Laplacian, i.e., it satisfies equation (4.59). Note that, in other textbooks, $\mathcal{G}$ is represented as $1/\nabla^2$.

What we have learnt with equation (4.87) is that standard canonical quantization is incompatible with quantization with constraints. In our case, we have used the Coulomb gauge condition and Gauss law as constraints, and these require a modification of the canonical commutators to the expression in equation (4.87). On the other hand, standard canonical quantization has been applied to the $a_{\mathbf{k},\lambda}$ and $a^{\dagger}_{\mathbf{k},\lambda}$ operators, in the sense that we can define a set of self-adjoint operators (coordinates) $q_{\mathbf{k},\lambda}$ and $p_{\mathbf{k},\lambda}$ (momenta), given by

$$a_{\mathbf{k},\lambda} = (2|\mathbf{k}|)^{-1/2}(|\mathbf{k}|q_{\mathbf{k},\lambda} + i\mathbf{p}_{\mathbf{k},\lambda}), \tag{4.88}$$

$$a^{\dagger}_{\mathbf{k},\lambda} = (2|\mathbf{k}|)^{-1/2}(|\mathbf{k}|q_{\mathbf{k},\lambda} - ip_{\mathbf{k},\lambda}), \tag{4.89}$$

which satisfy canonical commutation relations, and express the Hamiltonian as

$$H = \frac{1}{2}\sum_{\lambda=1,2} \int \frac{d\mathbf{k}}{(2\pi)^3}(p^2_{\mathbf{k},\lambda} + \mathbf{k}^2q^2_{\mathbf{k},\lambda}). \tag{4.90}$$

Taking all this together, we can represent the quantized Hamiltonian of the free electromagnetic field as

$$H = \sum_{\lambda=1,2} \int \frac{d\mathbf{k}}{(2\pi)^3}|\mathbf{k}|a^{\dagger}_{\mathbf{k},\lambda}a_{\mathbf{k},\lambda} + \mathcal{E}_0, \tag{4.91}$$

where $\mathcal{E}_0$ is an infinite zero-point energy that we formally write as

$$\mathcal{E}_0 = (2\pi)^3 \int \frac{d\mathbf{k}}{(2\pi)^3} |\mathbf{k}| \int \frac{d\mathbf{k}'}{(2\pi)^3} \delta^2(\mathbf{k} - \mathbf{k}'), \tag{4.92}$$

where $\delta^2(\mathbf{k})$ is the squared Dirac delta, which is nominally infinite, and is not a distribution in the sense of Schwartz.

4.2 Electrons in solids

A very important scenario where strong interactions play a major role is that of electrons in a crystalline solid. In this section, we provide the theoretical background necessary for later study of electronic interactions in a solid. We follow for the most part the structure in standard textbooks in solid state physics [4], that the reader is encouraged to consult for further details and much more information about the topic. We begin with the general study of crystalline arrangements, the formation of bands and band gaps, and the tight-binding approximation, all of which are essential for the rest of this book, and then introduce some low-energy effective models in the continuum that will be useful later on.

4.2.1 Crystal structures: basic concepts

Here, we will consider somewhat abstract mathematical structures given by sets in $\mathbb{R}^D$, which satisfy certain requirements. In particular, we introduce Bravais lattices, and related concepts, such as the first Brillouin zone. The first Brillouin zone is perhaps the most important concept for our purposes, and should not be overlooked.

We begin by considering D vectors $\mathbf{a}_i$ in $\mathbb{R}^D$, with the additional condition that they are not all coplanar. In one spatial dimension, this is trivial since we can pick any constant $a_1 \in \mathbb{R}$. For $D = 2$, coplanarity translates to collinearity, which simply means that the set $A_2 = \{\mathbf{a}_1, \mathbf{a}_2\}$ ($\mathbf{a}_i \in \mathbb{R}^2$) is linearly independent. In three dimensions, the set $A_3 = \{\mathbf{a}_1, \mathbf{a}_2, \mathbf{a}_3\}$ ($\mathbf{a}_i \in \mathbb{R}^3$) must be linearly independent and satisfy

$$\mathbf{a}_i \cdot (\mathbf{a}_j \times \mathbf{a}_k) \neq 0, \ i \neq j \neq k. \tag{4.93}$$

A set $S_D \subseteq \mathbb{R}^D$ is called a Bravais lattice with primitive vectors $\{\mathbf{a}_i\}_{i=1}^D$, not all coplanar, if the set of vectors of the form $\mathbf{R}(\mathbf{n})$, given by

$$\mathbf{R}(\mathbf{n}) = \sum_{i=1}^{D} n_i \mathbf{a}_i, \tag{4.94}$$

with $\mathbf{n} = (n_1, \dots, n_D) \in \mathbb{Z}^D$ equals S_D. Note that all Bravais lattices, as defined above, include the origin. This is obviously not a restriction, since we may as well displace all vectors of the lattice by a constant vector $\mathbf{a}'$ and still consider a (displaced) Bravais lattice. We shall stick to the original definition, with $\mathbf{a}' = 0$, for simplicity.

In one dimension, it is already clear that there is only one type of Bravais lattice, which is the linear crystal. All the points in S_1 are given by na, with n an integer and

a a constant, which we call the lattice spacing. In two dimensions, the classification of Bravais lattices is slightly richer, given the freedom to choose non-collinear vectors on the plane. To classify them, we may choose, without loss of generality, $\mathbf{a}_1 = \hat{x}$, and $\mathbf{a}_2$ a vector of length $|\mathbf{a}| \equiv a\ (>1)$ and at an angle $\theta\ (\in(0, \pi))$ with $\mathbf{a}_1$. This generates a family of two-dimensional Bravais lattices $S_2(a, \theta)$. In three dimensions, there are extra degrees of freedom, namely one more scale of length and another angle which, however, is constrained further by equation (4.93).

Let us now study several examples in $D = 2$. Choosing $a = 1$ and $\theta = \pi/2$, the Bravais lattice $S_2(1, \pi/2)$ represents a square lattice. For $a \neq 1$, $S_2(a, \pi/2)$ is called rectangular lattice (which may be simple or centred). If instead $\theta = 2\pi/3$, with $a = 1$, the lattice is called triangular or hexagonal. All other cases correspond to oblique lattices. This classification into five different types of two-dimensional Bravais lattices, which may look artificial, is actually quite important, and has to do with symmetry, i.e., group theory. The four symmetry groups (rectangular lattices come in two types) are C_2 (oblique), D_2 (rectangular), D_4 (square) and D_6 (hexagonal or triangular). Let us work out the example of the C_2 group, the two-dimensional, i.e., two-element, cyclic group. Having only two elements, one of them must be the identity e, and the other one, which we call c, satisfies $cc \equiv c^2 = e$. While the group admits a one-dimensional representation, with $e = 1$ and $c = -1$, the relevant representation for a two-dimensional lattice is two-dimensional. A valid representation is given by the 2×2 identity matrix and the parity operator $c = P$ (a π-rotation around the centre), with

$$P = \begin{pmatrix} -1 & 0 \\ 0 & -1 \end{pmatrix}, \tag{4.95}$$

which takes a vector $\mathbf{R} \in S_2(a, \theta)$ into $-\mathbf{R} \in S_2(a, \theta)$. This is the only symmetry held by an oblique lattice, as specified above. We now take on the symmetry group D_2 of the rectangular lattice, which has four elements, the identity e and $\{c_i\}_{i=1}^{3}$. The elements c_i correspond to a reflection about the vertical (c_1) and horizontal (c_2) axes, together with the parity operator ($c_3 = P$). The two reflections can be represented as the following 2×2 matrices

$$c_1 = \begin{pmatrix} 1 & 0 \\ 0 & -1 \end{pmatrix}, \quad c_2 = \begin{pmatrix} -1 & 0 \\ 0 & 1 \end{pmatrix}. \tag{4.96}$$

We encourage the reader to go through the same procedure for the higher-symmetry groups of the square (D_4) and hexagonal (D_6) lattices.

In three dimensions, the situation is much richer. We will consider the cubic lattices (simple, face-centred and body-centred), and study in more detail the simple cubic one, since this is important for partial-wave scattering in a finite, cubic volume. For the simple cubic lattice, we can choose three mutually orthogonal primitive vectors of equal length, i.e., $|\mathbf{a}_i| = a$ for $i = 1, 2, 3$ and $\mathbf{a}_i \cdot \mathbf{a}_j = a^2\delta_{i,j}$. Obviously, we identify a with the lattice spacing. For the body-centred cubic (bbc) lattice, we may choose $\mathbf{a}_1 = a\hat{x}$, $\mathbf{a}_2 = a\hat{y}$ and $\mathbf{a}_3 = (a/2)(\hat{x} + \hat{y} + \hat{z})$. These three vectors clearly generate the lattice, are linearly independent, and satisfy $\mathbf{a}_3 \cdot (\mathbf{a}_1 \times \mathbf{a}_2) = a^3/3 \neq 0$.

As for the face-centred cubic (fcc) lattice, one may choose $\mathbf{a}_1 = (a/2)(\hat{y} + \hat{z})$, $\mathbf{a}_2 = (a/2)(\hat{x} + \hat{z})$ and $\mathbf{a}_3 = (a/2)(\hat{x} + \hat{y})$, which is again linearly independent and not coplanar.

A useful concept in a Bravais lattice is the coordination number, which we shall denote z when there is no ambiguity. The coordination number is defined as the number of nearest neighbours that each lattice point has, which is obviously identical for every lattice point. As we shall see in section 4.2.3, this is a very useful concept in the tight-binding approximation for an electron in a periodic potential.

Another very useful, and needed concept is that of a lattice with a basis. Before introducing its definition, let us take a look at a periodic potential V in $D = 1$, given by

$$V(x) = V_1(x) + V_2(x), \tag{4.97}$$

with $V_1(x + b) = V_1(x)$ and $V_2(x + 2b) = V_2(x)$. Clearly, V is periodic with periodicity $2b$. Consider now the set of points $S = \{x\colon V(x) \text{ is minimum}\}$. Clearly, $S = \{na + x_0\}_{n\in\mathbb{Z}}$, being (arbitrarily) x_0 the smallest positive position of the minimum of $V(x)$, and $a = 2b$. After a displacement of all points in the set S by x_0 to the left, we are left with $S_1(a)$, that is, the one-dimensional Bravais lattice with lattice spacing a. Typically, a periodic potential of the type in equation (4.97), which is called a superlattice, or generally any finite superposition of periodic potentials with commensurate periods (i.e., their ratios are rational numbers), is best treated in physical situations by considering the points of the superlattice, which is not a Bravais lattice. In our example, the superlattice is the set $S_1^s(a, 2a) = S[V_1] \cup S[V_2]$, where $S[v] = \{x\colon v(x) \text{ is minimum}\}$. In that way, nearest neighbour points with respect to a point in $S[V_i]$ are defined as the closest points in $S_1^s(a, 2a)$. Clearly, points in the superlattice are not equivalent to all others, since it is not a Bravais lattice. We now introduce the concept of lattice with a basis, which, among other things, is important for superlattices.

Among other equivalent definitions, we define a lattice with an n-atom basis as follows. Consider a set A_D^n of n points $\mathbf{r}_i$ $(i = 1, \ldots n)$ in $\mathbb{R}^D$, including the origin. Consider a Bravais lattice $S_D \subseteq \mathbb{R}^D$ with primitive vectors $\{\mathbf{a}_i\}_{i=1}^D$. A lattice with a basis is the set of n Bravais lattices S_D and those (displaced) Bravais lattices generated by displacing S_D by every point $\mathbf{r}_i \in A_n^D$. When we consider lattices with a basis, we will simply define the Bravais lattice S_D and the n-atom set A_n^D.

Note that while generally the union of all Bravais lattices associated with a lattice with a basis is not a Bravais lattice, there are Bravais lattices that can be described as a lattice with a basis. This is the case of the bcc and fcc lattices, which can be described as simple cubic lattices with n-atom bases ($n = 2$ for bcc and $n = 4$ for fcc). Examples of lattices with a basis in which the union of all associated Bravais lattices are not Bravais lattices are easy to construct. For the one-dimensional superlattice described above, $S_1^s(a, 2a)$ is clearly not a Bravais lattice (recall there is only one Bravais lattice for $D = 1$). The honeycomb lattice ($D = 2$), realized in graphene, is also a lattice with a basis with this property. Since we shall consider the example of electrons in graphene later on, we elaborate here on the honeycomb lattice, and specify the Bravais lattice and the 2-atom set. The primitive vectors $\mathbf{a}_1$ and $\mathbf{a}_2$, basis vectors $\mathbf{r}_1$ and $\mathbf{r}_2$ and the three nearest-neighbour vectors ℓ_1, ℓ_2 and ℓ_3 are given by

$$\mathbf{a}_1 = a\hat{x},\ \mathbf{a}_2 = (a/2)(\hat{x} + \sqrt{3}\,\hat{y}), \tag{4.98}$$

$$\mathbf{r}_1 = 0,\ \mathbf{r}_2 = (a/\sqrt{3})\hat{y}, \tag{4.99}$$

$$\ell_1 = (a/2)(\hat{x} + \sqrt{3}\,\hat{y}),\ \ell_2 = -(a/\sqrt{3})\hat{y},\ \ell_3 = (a/2)(-\hat{x} + \sqrt{3}\,\hat{y}). \tag{4.100}$$

We continue our review of pertinent concepts now by introducing the primitive unit cell. Consider the point $\mathbf{R} = 0 \in \mathbb{R}^D$ of a Bravais lattice in D-dimensions with primitive vectors $\{\mathbf{a}_i\}_{i=1}^D$, and a closed (hyper)volume $V_D(0) \subset \mathbb{R}^D$ so that $0 \in V_D(0)\backslash\partial V_D(0)$ (i.e., 0 is an interior point), and no other point of the Bravais lattice belongs to $V_D(0)$. Define now $V_D(\mathbf{n}) = \{\mathbf{r} \in \mathbb{R}^D\colon \mathbf{r} = \mathbf{r}' + \sum_{i=1}^D n_i\mathbf{a}_i,\ \forall\,\mathbf{r}' \in V_D(0)\}$, with $\mathbf{n} \in \mathbb{Z}^D$. We say that $V_D(0)$ is a primitive unit cell if $\bigcup_{\mathbf{n}\in\mathbb{Z}^D} V_D(\mathbf{n}) = \mathbb{R}^D$ and $\bigcap_{\mathbf{n}\in\mathbb{Z}^D}[V_D(\mathbf{n})\backslash\partial V_D(\mathbf{n})] = \varnothing$. The most important type of primitive unit cell is the so-called Wigner–Seitz cell. This is defined as a primitive unit cell $V_D(0)$ such that every point in $\partial V_D(0)$ is closer to $\mathbf{R} = 0$ than to any other point in the Bravais lattice.

So far, we have only studied the spatial structure of crystal lattices. However, since discrete translation invariance is present on a Bravais lattice, many quantum-mechanical applications may benefit from a momentum representation of sorts. This will be called the quasi-momentum representation. We begin by defining the reciprocal lattice. Given a Bravais lattice with vectors $\mathbf{R}$, the reciprocal vector is the set containing all vectors $\mathbf{K}$ such that

$$e^{i\mathbf{K}\cdot\mathbf{R}} = 1. \tag{4.101}$$

The reciprocal lattice is itself a Bravais lattice, which we prove now. The prospective reciprocal lattice primitive vectors will be denoted $\mathbf{b}_i$. The original (also called direct) lattice has primitive vectors $\{\mathbf{a}_i\}_{i=1}^D$. In one dimension, obtaining the reciprocal lattice is trivial. Since $a_1 = 1$, then using equation (4.101), we have $b_1 = 2\pi$ (or $2\pi/a$ if we restore the lattice spacing). In two and three dimensions we choose (recall primitive vectors are not unique) the primitive vectors of the reciprocal lattice $\mathbf{b}_i$ such that $\mathbf{b}_i \cdot \mathbf{a}_j = 2\pi\delta_{i,j}$. For $D = 2$, we can rotate $\mathbf{a}_j$ by $\pi/2$ using the rotation matrix $\hat{\mathcal{R}}_{\pi/2}$ on $\mathbf{a}_j$ ($j \neq i$), with

$$\hat{\mathcal{R}}_\theta = \begin{pmatrix} \cos\theta & -\sin\theta \\ \sin\theta & \cos\theta \end{pmatrix}. \tag{4.102}$$

We then write $\mathbf{b}_i = \alpha_i\hat{\mathcal{R}}_{\pi/2}\mathbf{a}_j$ ($j \neq i$), with α_i two constants to be determined. Imposing the conditions above, and using our usual convention $\mathbf{a}_1 = a\hat{x}$, and $\mathbf{a}_2 = a_2^x\hat{x} + a_2^y\hat{y}$, we obtain

$$\mathbf{b}_1 = \frac{2\pi}{a}\left(\hat{x} - \frac{a_2^x}{a_2^y}\hat{y}\right),\ \mathbf{b}_2 = \frac{2\pi}{a_2^y}\hat{y}. \tag{4.103}$$

Note how $\mathbf{b}_1$ and $\mathbf{b}_2$ generate a two-dimensional Bravais lattice, since they are linearly independent. The three-dimensional case is similar, and standard methods give us

$$\mathbf{b}_i = 2\pi \frac{\mathbf{a}_{i+1} \times \mathbf{a}_{i+2}}{\mathbf{a}_1 \cdot (\mathbf{a}_2 \times \mathbf{a}_3)}, \tag{4.104}$$

where the indices $i + 1$ and $i + 2$ are defined cyclically (modulo 3), i.e., $4 \to 1$ and $5 \to 2$.

Let us analyse the two-dimensional example of the honeycomb lattice or, more concretely, its associated Bravais lattice, i.e., the triangular lattice. Introducing equation (4.98) into equation (4.104), we obtain

$$\mathbf{b}_1 = \frac{2\pi}{a}\left(\hat{x} - \frac{1}{\sqrt{3}}\hat{y}\right), \mathbf{b}_2 = \frac{4\pi}{\sqrt{3}a}\hat{y}. \tag{4.105}$$

To see clearly the geometry of the reciprocal lattice, we can define $b \equiv 4\pi/\sqrt{3}a$, so that equation (4.105) is written as

$$\mathbf{b}_1 = \frac{b}{2}(-\hat{y} + \sqrt{3}\hat{x}), \mathbf{b}_2 = b\hat{y}, \tag{4.106}$$

which is obviously another triangular (hexagonal) lattice, rotated by $\pi/2$, which is the case by construction.

The last definition we will need is that of the first Brillouin zone. This is simply the Wigner–Seitz cell of the reciprocal lattice. It will appear constantly when dealing with particles in lattices, and as a domain integration, it will be denoted by either BZ or 1BZ.

4.2.2 Band structure

One of the most well-known and important facts of non-relativistic particles subjected to a periodic potential is the emergence of energy bands and forbidden energies in the spectrum (called gaps). A free non-relativistic quantum particle has the simple Hamiltonian $H = \mathbf{p}^2/2m$, which is translation invariant. All (non-normalizable) eigenstates are given by plane waves $\mathbf{k}$ with energies $\hbar^2\mathbf{k}^2/2m$ which, in the position representation are simply

$$\langle \mathbf{r}|\mathbf{k}\rangle = e^{i\mathbf{k}\cdot\mathbf{r}}, \tag{4.107}$$

normalized as $\langle \mathbf{k}'|\mathbf{k}\rangle = (2\pi)^D\delta(\mathbf{k} - \mathbf{k}')$. A translation by a vector $\mathbf{r}'$ is realized by an operator $T(\mathbf{r}')$ (the translation operator by a vector $\mathbf{r}'$) which acts on position eigenstates $\mathbf{r}$ as

$$T(\mathbf{r}')|\mathbf{r}\rangle = |\mathbf{r} + \mathbf{r}'\rangle. \tag{4.108}$$

Hence, its action on a plane wave is

$$
\begin{aligned}
T(\mathbf{r}')|\mathbf{k}\rangle &= \int d\mathbf{r} T(\mathbf{r}')|\mathbf{r}\rangle\langle\mathbf{r}|\mathbf{k}\rangle \\
&= \int d\mathbf{r}|\mathbf{r} + \mathbf{r}'\rangle\langle\mathbf{r}|\mathbf{k}\rangle \\
&= \int d\mathbf{r}|\mathbf{r}\rangle\langle\mathbf{r} - \mathbf{r}'|\mathbf{k}\rangle \\
&= e^{-i\mathbf{k}\cdot\mathbf{r}'}|\mathbf{k}\rangle,
\end{aligned}
\tag{4.109}
$$

which leaves the physics unchanged, because it only adds a constant phase $(-\mathbf{k} \cdot \mathbf{r}')$ to the state $|\mathbf{k}\rangle$. Its adjoint $T^{\dagger}(\mathbf{r}')$ is also a translation operator, since

$$
T^{\dagger}(\mathbf{r}')|\mathbf{r}\rangle = |\mathbf{r} - \mathbf{r}'\rangle, \tag{4.110}
$$

that is, $T^{\dagger}(\mathbf{r}') = T(-\mathbf{r}')$. Obviously, the composition, or product, of two translations $T(\mathbf{r}_2)T(\mathbf{r}_1)|\mathbf{r}\rangle = T(\mathbf{r}_2)|\mathbf{r} + \mathbf{r}_1\rangle = |\mathbf{r} + \mathbf{r}_1 + \mathbf{r}_2\rangle$ and, therefore, we can write as their product rule

$$
T(\mathbf{r}_1)T(\mathbf{r}_2) = T(\mathbf{r}_1 + \mathbf{r}_2), \tag{4.111}
$$

which is also a translation operator. The operator product is associative, there is an identity translation, which is $T(0) = 1$, and every translation operator has an inverse, given by its adjoint. Hence, the set of all translations form a group, which is also trivially seen to be Abelian. Also, since $T^{-1}(\mathbf{r}') = T^{\dagger}(\mathbf{r})$, translation operators are unitary. The group they conform is called translation group. In the end, what we are most interested in is the fact that $[H, T(\mathbf{r}')] = 0 \; \forall \mathbf{r}'$, so that the eigenstates of H can be chosen to be simultaneously as eigenstates of $T(\mathbf{r}')$ for all $\mathbf{r}'$. This is exactly what we did at the start, that is, choose as a basis of eigenstates the set of all plane waves $|\mathbf{k}\rangle$, which are also eigenstates of $T(\mathbf{r}')$ with eigenvalue $\exp(-i\mathbf{k} \cdot \mathbf{r}')$, see equation (4.109).

For particles in a crystalline medium, whether electrons in a solid or atoms in an optical lattice, we are interested in a subgroup of the translation group, since the single-particle Hamiltonian is not translationally invariant, but instead has discrete translation invariance. Let us write down a typical single-particle Hamiltonian H

$$
H = \frac{\mathbf{p}^2}{2m} + V(\mathbf{r}), \tag{4.112}
$$

and a Bravais lattice S_D with primitive vectors $\mathbf{a}_i$ $(i = 1, \dots, D)$. We assume that $V(\mathbf{r} + \mathbf{R}) = V(\mathbf{r})$ for all $\mathbf{R} \in S_D$. Now, for a given $\mathbf{R}$ of the Bravais lattice, the corresponding translation operator $T(\mathbf{R})$ acts on a position eigenstate $|\mathbf{r}\rangle$ as in equation (4.108), with $\mathbf{r}'$ replaced with $\mathbf{R}$. For two given vectors of the lattice, $\mathbf{R}$ and $\mathbf{R}'$, obviously $T(\mathbf{R})T(\mathbf{R}') = T(\mathbf{R}')T(\mathbf{R}) = T(\mathbf{R} + \mathbf{R}')$ is also a translation operator. And since $\mathbf{R} + \mathbf{R}' \in S_D$, the set of all translation operators $\mathcal{T}(S_D) \equiv \{T(\mathbf{R}): \mathbf{R} \in S_D\}$ forms a group, and therefore $\mathcal{T}(S_D)$ is a subgroup of the translation group. The kinetic energy operator $\mathbf{p}^2/2m$ commutes with all elements of $\mathcal{T}(S_D)$, and so does the potential, since

$$
\begin{aligned}
T(\mathbf{R})V &= \int d\mathbf{r}T(\mathbf{R})|\mathbf{r}\rangle V(\mathbf{r})\langle\mathbf{r}| = \int d\mathbf{r}|\mathbf{r}+\mathbf{R}\rangle V(\mathbf{r})\langle\mathbf{r}| \\
&= \int d\mathbf{r}|\mathbf{r}\rangle V(\mathbf{r}-\mathbf{R})\langle\mathbf{r}-\mathbf{R}| = \int |\mathbf{r}\rangle V(\mathbf{r})(T^{\dagger}(\mathbf{R})|\mathbf{r}\rangle)^{\dagger} = VT(\mathbf{R}).
\end{aligned}
\tag{4.113}
$$

We now look for the general form of the eigenstates of $T(\mathbf{R})$. Obviously, plane waves are eigenstates (see equation (4.109)). Also, any function $\phi(\mathbf{r})$ that has the periodicity of the Bravais lattice, i.e., $\phi(\mathbf{r}+\mathbf{R}) = \phi(\mathbf{r})$ is also an eigenstate of all $T(\mathbf{R})$ with eigenvalue 1 ($T(\mathbf{R})|\phi\rangle = |\phi\rangle$). Therefore, any wave function $|\psi_{\mathbf{k}}\rangle$ with position representation

$$
\langle\mathbf{r}|\psi_{\mathbf{k}}\rangle = \phi_{\mathbf{k}}(\mathbf{r})e^{i\mathbf{k}\cdot\mathbf{r}}, \tag{4.114}
$$

with $\phi_{\mathbf{k}}(\mathbf{r}+\mathbf{R}) = \phi_{\mathbf{k}}(\mathbf{r})$ is an eigenstate of all $T(\mathbf{R})$, with

$$
T(\mathbf{R})|\psi_{\mathbf{k}}\rangle = e^{-i\mathbf{k}\cdot\mathbf{R}}|\psi_{\mathbf{k}}\rangle. \tag{4.115}
$$

Clearly, the eigenvalues $t_{\mathbf{k}} = \exp(-i\mathbf{k}\cdot\mathbf{R})$ have the periodicity of the reciprocal lattice. When looking for common eigenstates of H and $T(\mathbf{R})$, we can restrict the values of $\mathbf{k}$ to the first Brillouin zone.

We may now conclude by stating Bloch's theorem: every eigenstate of Hamiltonian H in (4.112) such that $H = \mathcal{T}^{-1}(S_D)H\mathcal{T}(S_D)$ is of the form of equation (4.114), with $\phi_{\mathbf{k}}(\mathbf{r})$ periodic with the periodicity of S_D, and all eigenenergies are periodic in $\mathbf{k}$ with the periodicity of the first Brillouin zone.

An important property of particles trapped in periodic potentials is the opening of 'band gaps', that is, forbidden energy regions, due to coupling between otherwise degenerate states due to the periodic potential. The simplest way to see this is by considering a single particle in one spatial dimension subjected to a periodic potential $V(x)$ such that $V(x+a) = V(x)$. In the non-interacting (plane wave) basis in a ring of length $L = Na$, with N an arbitrary integer, we have the following matrix elements

$$
\langle k'|V|k\rangle = \frac{1}{L}\int_0^L dx e^{i(k-k')x}V(x) = \frac{1}{Na}\int_0^{Na} dx e^{i2\pi(n-n')x/Na}V(x). \tag{4.116}
$$

Since the potential is periodic with period a, it admits a Fourier series decomposition of the form

$$
V(x) = \sum_{n=-\infty}^{\infty} v_n e^{i2\pi nx/a}. \tag{4.117}
$$

Inserting the Fourier expansion into equation (4.116), we obtain

$$
\langle k'|V|k\rangle = \frac{1}{Na}\sum_{m\in\mathbb{Z}}\int_0^{Na} dx \exp\left[i2\pi x((n-n')/N + m)/a\right]. \tag{4.118}
$$

As $L = Na \to \infty$, the only non-zero matrix elements correspond to values of $k - k'$ that are multiples of the reciprocal lattice vector $q = 2\pi/a$. For the lowest band, this corresponds to $k - k' = 2\pi/a$. The two states that are degenerate in the non-interacting limit and satisfy this condition are the points at the edges of the Brillouin zone, that is, $k = \pm\pi/a \equiv \kappa$. Applying degenerate perturbation theory to these two states, we are left with the diagonalization of the following matrix

$$H = \begin{pmatrix} \frac{\hbar^2\kappa^2}{2m} & W \\ W^* & \frac{\hbar^2\kappa^2}{2m} \end{pmatrix}, \tag{4.119}$$

where W is the matrix element of V between the states $|\pm\kappa\rangle$. Therefore, the two energies split as

$$E_{\pm} = \frac{\hbar^2\kappa^2}{2m} \pm |W|, \tag{4.120}$$

and, clearly, the band gap becomes, perturbatively, $\Delta = 2|W|$.

4.2.3 Tight-binding approximation

In this section we consider a very important method for calculating band structures in certain limits. Moreover, introducing interactions within the tight-binding method is simpler than in the full-blown theory, and the method connects nicely with lattice effective field theory (l-EFT) [5].

We begin by considering, as usual, a Bravais lattice S_D in D spatial dimensions, and a single-particle, non-relativistic, periodic Hamiltonian with the periodicity of S_D, given by $H = H_0 + V$. The potential V, when dealing with an electron in a solid, is generated by the crystallized ionic structure of the material. If the crystal is atomic, then we may represent each point of the Bravais lattice as the equilibrium position of each atom within the crystal. The lattice spacing or, more concretely, the distance between nearest-neighbours is assumed to be large when compared with a certain localization length ξ that we define now. We imagine the electron, which in principle traverses the lattice, interacting with a single, isolated core ion. The ion is assumed static ($M/m_e \gg 1$, with M and m_e the atomic and electronic masses, respectively) and the electronic Hamiltonian is called $\tilde{H} = H_0 + \tilde{V}$, for an ion localized at $\mathbf{r} = 0$. We obtain bound states $|\psi_n\rangle$ of $\tilde{H}$ with eigenenergies E_n, and assume their localization length ξ is much smaller than the lattice spacing a. In that case, the eigenstates $\psi_n(\mathbf{r}) \approx 0$ for $|\mathbf{r}| > a$.

Let us write the full periodic Hamiltonian as

$$H = H_0 + V = H_0 + \tilde{V} + (V - \tilde{V}) \equiv \tilde{H} + \Delta V. \tag{4.121}$$

Since ΔV has very little overlap with the atomic eigenstates, a reasonably good approximation to the *eigenvalues* is obtained E_n. However, H in equation (4.121) is periodic, and therefore Bloch's theorem tells us that the eigenstates must have the Bloch form in equation (4.114). A so-called linear combination of atomic orbitals (LCAO), would be a good ansatz for the eigenstates, with the form

$$\psi_{n\mathbf{k}}(\mathbf{r}) = \sum_{\mathbf{R}\in S_D} e^{i\mathbf{k}\cdot\mathbf{R}}\psi_n(\mathbf{r}-\mathbf{R}). \tag{4.122}$$

Clearly, $\psi_{n\mathbf{k}}$ above is of the Bloch form (4.114), and is an admissible wave function. The above ansatz, however, is not exact. We propose instead an eigenfunction expansion of the form

$$\psi_{n\mathbf{k}}(\mathbf{r}) = \sum_{\mathbf{R}\in S_D} e^{i\mathbf{k}\cdot\mathbf{R}}w_n(\mathbf{r}-\mathbf{R}), \tag{4.123}$$

where w_n is a so-called Wannier function (corresponding to band index n).

The expansion (4.123) is general, since Wannier functions form an orthogonal basis. To see this, let us assume that $\psi_{n\mathbf{k}}$ in equation (4.123) is a Bloch eigenstate of H. We multiply equation (4.123) by $\exp(-i\mathbf{k}\cdot\mathbf{R})/\Omega$, with $\mathbf{R}\in S_D$, and Ω the volume of the first Brillouin zone, and integrate over $\mathbf{k}\in \mathrm{BZ}$, obtaining

$$\int_{\mathrm{BZ}}\frac{d\mathbf{k}}{\Omega}e^{-i\mathbf{k}\cdot\mathbf{R}}\psi_{n\mathbf{k}}(\mathbf{r}) = \sum_{\mathbf{R}'\in S_D} w_n(\mathbf{r}-\mathbf{R}')\int_{\mathrm{BZ}}\frac{d\mathbf{k}}{\Omega}e^{i\mathbf{k}\cdot(\mathbf{R}'-\mathbf{R})} = w_n(\mathbf{r}-\mathbf{R}). \tag{4.124}$$

In the last equality above, we have used the fact that

$$\int_{\mathrm{BZ}}\frac{d\mathbf{k}}{\Omega}e^{i\mathbf{k}\cdot(\mathbf{R}'-\mathbf{R})} = \delta_{\mathbf{R}',\mathbf{R}}. \tag{4.125}$$

We show now that Wannier functions corresponding to different Bravais lattice vectors $\mathbf{R}$ and $\mathbf{R}'$ are orthogonal, and so are Wannier functions corresponding to different band indices n and n'. To do this, let us calculate the scalar product $\langle w_{n'}(\mathbf{R}')|w_n(\mathbf{R})\rangle$, with $\langle\mathbf{r}|w_n(\mathbf{R})\rangle = w_n(\mathbf{r}-\mathbf{R})$, as

$$\int d\mathbf{r}w_{n'}^*(\mathbf{r}-\mathbf{R}')w_n(\mathbf{r}-\mathbf{R}) = \int_{\mathrm{BZ}}\frac{d\mathbf{k}'}{\Omega}\int_{\mathrm{BZ}}\frac{d\mathbf{k}}{\Omega}e^{i(\mathbf{k}'\cdot\mathbf{R}'-\mathbf{k}\cdot\mathbf{R})}\int d\mathbf{r}\psi_{n'\mathbf{k}'}^*(\mathbf{r})\psi_{n\mathbf{k}}(\mathbf{r}). \tag{4.126}$$

It is easy to see that the scalar product of two Bloch functions is calculated as

$$\int d\mathbf{r}\psi_{n'\mathbf{k}'}^*(\mathbf{r})\psi_{n\mathbf{k}}(\mathbf{r}) = \delta_{n,n'}\delta_{\mathrm{BZ}}(\mathbf{k}-\mathbf{k}'), \tag{4.127}$$

where $\delta_{\mathrm{BZ}}(\mathbf{k}) = \sum_{\mathbf{K}\in\tilde{S}_D}\delta(\mathbf{k}-\mathbf{K})$ is the periodic delta function (or delta function mod BZ). Hence, equation (4.126) gives

$$\langle w_{n'}(\mathbf{R}')|w_n(\mathbf{R})\rangle = \delta_{n,n'}\int_{\mathrm{BZ}}\frac{d\mathbf{k}}{\Omega}e^{i\mathbf{k}\cdot(\mathbf{R}-\mathbf{R}')} = \delta_{n,n'}\delta_{\mathbf{R},\mathbf{R}'}. \tag{4.128}$$

For completeness, we prove that $\langle\mathbf{r}|w_n(\mathbf{R})\rangle = w_n(\mathbf{r}-\mathbf{R})$, which so far we have written without justification. It suffices to apply the translation operator $T(-\mathbf{R})$ on $|w_n(\mathbf{0})\rangle$. From equation (4.124), and using Bloch's theorem on $\psi_{n\mathbf{k}}$ we have

$$\langle\mathbf{r}|T(-\mathbf{R})|w_n(0)\rangle = \int_{\mathrm{BZ}}\frac{d\mathbf{k}}{\Omega}\psi_{n\mathbf{k}}(\mathbf{r}+\mathbf{R}) = \int_{\mathrm{BZ}}\frac{d\mathbf{k}}{\Omega}\psi_{n\mathbf{k}}(\mathbf{r})e^{i\mathbf{k}\cdot\mathbf{R}} = w_n(\mathbf{r}+\mathbf{R}), \tag{4.129}$$

that is, $|w_n(\mathbf{R}'-\mathbf{R})\rangle = T(-\mathbf{R})|w_n(\mathbf{R}')\rangle$. Since $|w_n(0)\rangle$ does not depend explicitly on $\mathbf{R}$, $\langle\mathbf{r}|w_n(\mathbf{R})\rangle = w_n(\mathbf{r}-\mathbf{R})$.

After the detour on Wannier functions, we are ready to implement the tight-binding approximation. Inserting the expansion of the Bloch eigenstates in terms of the Wannier functions, equation (4.123), into the stationary Schrödinger equation with Hamiltonian H, equation (4.121), and projecting onto atomic eigenstates (eigenstates of $\tilde{H}$), with eigenvalues E_m, we obtain

$$(E - E_m) \sum_{\mathbf{R} \in S_D} e^{i\mathbf{k}\cdot\mathbf{R}} \int d\mathbf{r} \psi_m^*(\mathbf{r}) w_n(\mathbf{r} - \mathbf{R}) = \sum_{\mathbf{R} \in S_D} e^{i\mathbf{k}\cdot\mathbf{R}} \Delta V(\mathbf{r}) \psi_m^*(\mathbf{r}) w_n(\mathbf{r} - \mathbf{R}). \tag{4.130}$$

We further expand the Wannier function $|w_n(\mathbf{R})\rangle$ in terms of atomic eigenstates as

$$w_n(\mathbf{r} - \mathbf{R}) = \sum_j b_j^{(n)} \psi_j(\mathbf{r} - \mathbf{R}), \tag{4.131}$$

and inserting this into equation (4.130) and rearranging terms, we obtain

$$\begin{aligned} &(E - E_m) b_m^{(n)} + (E - E_m) \sum_{\mathbf{R} \neq 0} e^{i\mathbf{k}\cdot\mathbf{R}} \sum_j b_j^{(n)} \int d\mathbf{r} \psi_m^*(\mathbf{r}) \psi_j(\mathbf{r} - \mathbf{R}) \\ &= \sum_j b_j^{(n)} \int d\mathbf{r} \psi_m^*(\mathbf{r}) \Delta V(\mathbf{r}) \psi_j(\mathbf{r}) + \sum_{\mathbf{R} \neq 0} \sum_j b_j^{(n)} e^{i\mathbf{k}\cdot\mathbf{R}} \int d\mathbf{r} \Delta V(\mathbf{r}) \psi_m^*(\mathbf{r}) \psi_j(\mathbf{r} - \mathbf{R}). \end{aligned} \tag{4.132}$$

The awful-looking expression in equation (4.132) can actually be made useful. Let us consider a dominant s-level (angular momentum $\ell = 0$), with $b_j^{(n)} \approx \delta_{j,0}$, and $n \equiv 0$. Setting $E = \epsilon(\mathbf{k})$, equation (4.132) simplifies to

$$\epsilon(\mathbf{k}) - E_0 \approx \frac{\int d\mathbf{r} \Delta V(\mathbf{r}) \left[|\psi_0(\mathbf{r})|^2 + \sum_{\mathbf{R} \neq 0} e^{i\mathbf{k}\cdot\mathbf{R}} \psi_0^*(\mathbf{r}) \psi_0(\mathbf{r} - \mathbf{R}) \right]}{1 + \sum_{\mathbf{R} \neq 0} \int d\mathbf{r} \psi_0^*(\mathbf{r}) \psi_0(\mathbf{r} - \mathbf{R}) e^{i\mathbf{k}\cdot\mathbf{R}}}. \tag{4.133}$$

We can approximately simplify equation (4.133) by assuming negligible overlap between atomic eigenstates further than nearest neighbours. Regrouping $\mathbf{k}$-independent constants into $\bar{E}_0$, we obtain the tight-binding s-level dispersion

$$\epsilon(\mathbf{k}) \approx \bar{E}_0 + \frac{A + \sum_{\langle \mathbf{R}, 0 \rangle} B(\mathbf{R}) e^{i\mathbf{k}\cdot\mathbf{R}}}{1 + \sum_{\langle \mathbf{R}, 0 \rangle} C(\mathbf{R}) e^{i\mathbf{k}\cdot\mathbf{R}}}. \tag{4.134}$$

Above A, B and C are easily read from equation (4.133), and the notation $\langle \mathbf{R}, \mathbf{0} \rangle$ denotes sum over $\mathbf{R} \in S_D$ that are nearest neighbours with the origin.

4.2.4 Low-energy non-interacting theories

Let us study some tight-binding dispersions $\epsilon(\mathbf{k})$ in different dimensions that have interesting low-energy behaviour. Firstly, low-energy may mean truly single-particle low-energy, or low-energy with respect to a particularly relevant (vector) quasi-momentum, such as a point at the Fermi surface.

We begin with $D = 1$, which is by far simplest. The dispersion of the lowest band of a standard Bravais lattice, with no basis, is given, up to a constant, by

$$\epsilon(k) = -2J\cos(kd) + 2J, \tag{4.135}$$

where d is the lattice spacing, and k is in the first Brillouin zone $(-\pi/d, \pi/d]$. If we are interested in bosons, or very low-density systems at low temperature, then we may expand the dispersion in equation (4.135) around its minimum at $k = 0$. We have

$$\epsilon(k) = J(kd)^2 + O((kd)^4) \equiv \frac{\hbar^2 k^2}{2m_*} + O((kd)^4),\ |kd| \to 0. \tag{4.136}$$

That is, the dispersion is essentially non-relativistic with effective mass m_* ($J = \hbar^2/2m_* d^2$), as long as $|kd|$ is small. Another interesting limit is relevant for fermions at high densities. For $kd \approx \pm\pi/2$, we write $kd \equiv \pm\pi/2 + qd$. The dispersion (4.135), up to an irrelevant constant, can be expanded as

$$\epsilon(k) = 2J|qd| + O((qd)^3) \equiv \hbar v|q| + O((qd)^3),\ |qd| \to 0. \tag{4.137}$$

That is, the dispersion is approximately linear, with speed v ($J = \hbar v/2d$). A lattice with a basis—superlattice—is also interesting and provides more interesting low-energy dispersions. When the basis has two atoms, the dispersion, up to a constant, splits into two subbands, with energies $\epsilon_n(\mathbf{k})$, with $n = 1, 2$, given by

$$\epsilon_n(k) = (-1)^n\sqrt{[2J\cos(kd)]^2 + \lambda^2}. \tag{4.138}$$

The most interesting case now corresponds to $kd \approx \pm\pi/2$, with $kd = \pm\pi/2 + qd$, and the dispersion (4.138) is approximated by

$$\epsilon_n(k) \approx (-1)^n\sqrt{(2Jqd)^2 + \lambda^2} \equiv (-1)^n\sqrt{(mc^2)^2 + (\hbar qc)^2}, \tag{4.139}$$

which is the relativistic dispersion, where we have identified $\lambda = mc^2$ (half the mass gap), and $J = \hbar c/2d$. Note how this value of J coincides (identifying $v \leftrightarrow c$) with the one used for the gapless dispersion (4.135) near $\pm\pi/2$. This is natural, since that dispersion is obtained by setting $\lambda = 0$, that is, the massless case.

The two-dimensional case is more interesting, since geometry plays a bigger role. We briefly examine the particular case of the honeycomb lattice of graphene. The dispersion takes the form ($n = 1, 2$)

$$E_n(\mathbf{k}) = (-1)^n J\sqrt{3 + f(\mathbf{k})}, \tag{4.140}$$

where f is given by

$$f(\mathbf{k}) = 2\cos\left(\sqrt{3}k_y d\right) + 4\cos\left(\frac{\sqrt{3}}{2}k_y d\right)\cos\left(\frac{3}{2}k_x d\right). \tag{4.141}$$

In this case, the interesting and physically relevant expansion is done around the two so-called Dirac points $\mathbf{K}_{\pm}$ in the first Brillouin zone. These are given by

$$\mathbf{K}_{\pm} = \frac{2\pi}{3d}(\hat{x} \pm \hat{y}/\sqrt{3}). \tag{4.142}$$

What we do now is first expand f in powers of $k_x - 2\pi/3d$, and then in powers of $k_y \mp 2\pi/3\sqrt{3}d$. Clearly, the two subbands E_n touch at $\mathbf{k} = \mathbf{K}_{\pm}$, where $E_n(\mathbf{K}_{\pm}) = 0$, $n = 1, 2$, and are symmetric around these, which means the Fermi energy is located at zero energy. Doing the expansion, and introducing it back into equation (4.140) we obtain

$$E_n(\mathbf{K}_{\pm} + \mathbf{q}) \approx (-1)^n \frac{3Jd}{2}|\mathbf{q}| \equiv (-1)^n \hbar v|\mathbf{q}|. \tag{4.143}$$

Therefore, we have two (one around each Dirac point) massless relativistic dispersions.

In three dimensions, many more options open up. However, we shall not deal with them much in this book, except for the simple cubic lattice, which is completely analogous to the one-dimensional case, as it is separable.

4.3 Photons in structured media: cavities

Photons can exhibit a variety of interesting phenomena thanks to the current capabilities to enclose them in finite volumes (optical cavities), couple them with atoms, even couple cavity photons with atoms within the cavity, as well as implementing them in arrays of coupled cavities akin a periodic potential. We describe here a setting where photons can have remarkable effects and/or behaviour. In this section, we also set the speed of light in vacuum to unity ($c = 1$).

The most straightforward medium for photons is an optical cavity. An optical cavity can have many shapes and configurations. Typically, the simplest, ideal one consists of two perfectly reflecting mirrors, i.e., perfectly conducting plates, of zero thickness, planar, infinitely large, infinitely massive and rigid, and parallel to each other, located at a distance L from each other. This idealization corresponds to a one-dimensional cavity, since it constraints the fields only in one direction.

We take the electric field $\mathbf{E}$ in the $\hat{x}$-direction, that is

$$\mathbf{E}(\mathbf{r}, t) = \hat{x}E(z, t) \Rightarrow \mathbf{B}(\mathbf{r}, t) = \hat{y}B(z, t). \tag{4.144}$$

Perfect mirrors located at $x = 0$ and $x = L$ impose the following boundary conditions on the (classical) electric field:

$$E(z = 0, t) = E(z = L, t) = 0. \tag{4.145}$$

Since we are (in principle) only interested in the modes inside the cavity, we restrict $z \in [0, L]$, and the modes of the wave equation (4.67) take the form

$$A_n(z, t) = \sqrt{\frac{2}{L}}\sin(k_n z)e^{-ik_n t}, \; k_n = \frac{n\pi}{L}, \; n = 1, 2. \ldots \tag{4.146}$$

Therefore, the electric field, already quantized, takes the form $E = \sum_{n>0} E_n$, with

$$E_n(z, t) = \sqrt{\frac{k_n}{L}}\left[a_{k_n}(t) + a_{k_n}^{\dagger}(t)\right]\sin(k_n z), \tag{4.147}$$

where $a_{k_n}^{\dagger}(t) = e^{ik_n t} a_{k_n}^{\dagger}$.

We now derive a paradigmatic model of quantum optics, which describes the interaction of an atom with a near-resonant cavity photon: the Jaynes–Cummings model. Let us consider a (photonic) cavity mode with frequency ω ($=k_n$ for some n), and an atom placed inside the one-dimensional cavity at a fixed location z_0. The Hamiltonian H_A of the atom, when isolated, is given by equation (4.11), that is

$$H_A = \frac{1}{2}\hbar\omega_{eg}\sigma_z. \tag{4.148}$$

The Hamiltonian H_F of the field, assuming that its frequency is near-resonant with the transition frequency ω_{eg} of the atom, can be written in the single-mode approximation

$$H_F = \hbar\omega a^{\dagger} a, \tag{4.149}$$

where $a = a_{k_n}$, with $\omega = k_n$ the closest frequency to the atomic transition frequency. The coupling H_I between the atom and the field is obtained within the dipole approximation, i.e.

$$H_I = -dE(z_0), \tag{4.150}$$

which, upon quantization, equation (4.147) at $t = 0$, and using equation (4.14) gives

$$H_I = \sqrt{\frac{\omega}{L}}\,\sin(\omega z_0)(a + a^{\dagger})(v_{ge}\sigma_- + v_{ge}\sigma_+) \equiv \hbar g(\sigma_- + \sigma_+)(a + a^{\dagger}), \tag{4.151}$$

where we have set, without loss of generality, real $v_{eg} = v_{ge}$. We now pass to the interaction picture, with the non-interacting Hamiltonian $H_0 = H_A + H_F$. For $t_0 = 0$, the evolution operator $U_0(t, 0) \equiv U_0(t) = \exp(-iH_0 t/\hbar)$, and transform H_I as

$$\begin{aligned} H_I(t) = U_0^{\dagger}(t) H_I U_0(t) = \hbar g(&a^{\dagger}\sigma_- e^{i(\omega-\omega_{eg})t} + \sigma_+ a e^{-i(\omega-\omega_{eg})t} \\ &+ a^{\dagger}\sigma_+ e^{i(\omega+\omega_{eg})t} + \sigma_- a e^{-i(\omega+\omega_{eg})t}). \end{aligned} \tag{4.152}$$

Since we are only interested in near-resonant processes, $|\omega - \omega_{eg}| \ll \omega_{eg}$, we perform the so-called rotating-wave approximation (RWA), which consists of neglecting terms that oscillate with large frequencies $\omega + \omega_{eg}$, obtaining for the full Hamiltonian

$$H \approx \frac{1}{2}\hbar\omega_{eg}\sigma_z + \hbar\omega a^{\dagger} a + \hbar g(\sigma_+ a + a^{\dagger}\sigma_-). \tag{4.153}$$

The above Hamiltonian is called the Jaynes–Cummings model, which describes resonant atomic excitation (de-excitation) accompanied by photon absorption (emission).

References

[1] Pethick C J and Smith H 2008 *Bose-Einstein Condensation in Dilute Gases* (Cambridge: Cambridge University Press)

[2] Viebahn K, Sbroscia M, Carter E, Yu C Jr and Schneider U 2019 Matter-wave diffraction from a quasicrystalline optical lattice *Phys. Rev. Lett.* **122** 110404

[3] Dalibard J, Gerbier F, Juzeliunas G and Öhberg P 2011 Colloquium: artificial gauge potentials for neutral atoms *Rev. Mod. Phys.* **83** 1523

[4] Ashcroft N W and Mermin N D 2011 *Solid State Physics* (New York: Holt, Rinehart and Winston)

[5] Lähde T and Meißner U-G 2019 *Nuclear Lattice Effective Field Theory: An Introduction* (Berlin: Springer)

IOP Publishing

Strongly Interacting Quantum Systems, Volume 1

Few-body physics

Manuel Valiente and Nikolaj Thomas Zinner

Chapter 5

Dimensional reduction of trapped interacting quantum systems

In this chapter, we introduce the main techniques to effectively reduce the dimensions of a three-dimensional system to one spatial dimension via anisotropic trapping. We shall deal with scattering and bound states using two different methods, after introducing one-dimensional scattering theory, which is necessary for the correct interpretation of the results.

5.1 Collision theory in one dimension

We study here pure one-dimensional scattering, partially following reference [1]. Instead of going through the time-dependent formalism, we will directly study two-body, non-relativistic scattering with the Schrödinger equation, and also in the momentum representation with the Lippmann–Schwinger equation.

The stationary Schrödinger equation for two identical particles of mass m interacting via a local potential $V(z)$ is given by

$$-\frac{\hbar^2}{m}\psi''(z) + V(z)\psi(z) = E\psi(z). \tag{5.1}$$

Newton's third law implies that $V(z) = V(-z)$, that is, $V = V(|z|)$. Therefore, the Hamiltonian $H = H_0 + V$ commutes with the parity operator $\hat{P}$, that is, $[H, \hat{P}] = 0$. In the one-dimensional case, $\hat{P}$ plays the role of the angular momentum, and there are only two allowed values of this, namely $L = 0$ and $L = 1$, corresponding to the two eigenvalues p of $\hat{P}$, $p = \pm 1$. Even parity states ($p = +1$ or $L = 0$) constitute the 'even waves', while odd parity states ($p = -1$ or $L = 1$) constitute the 'odd waves'. These are the only two partial waves in one-dimensional scattering. It will be very

doi:10.1088/978-0-7503-3087-9ch5

useful to construct almost local parity eigenstates (ALPE), which we call $|Y_p^x\rangle$. These are constructed as

$$|Y_p^x\rangle \equiv \hat{\Pi}_p|x\rangle = A[|x\rangle + p|-x\rangle], \tag{5.2}$$

where A is a normalization constant such that $\hat{\Pi}_p^n|x\rangle = \hat{\Pi}_p|x\rangle$ for $n \in \mathbb{Z}_+$. By applying $\hat{\Pi}_p^2$ on $|x\rangle$, we see that $A = 1/2$, and we have

$$\langle x'|\hat{\Pi}_p|x\rangle = \frac{1}{2}\left[\delta(x - x') + p\delta(x + x')\right]. \tag{5.3}$$

The operator $\hat{\Pi}_p$ is also self-adjoint. Clearly, if we want to restrict an observable in position representation to a particular parity sector, we simply calculate their matrix elements between ALPEs (5.2). Given an observable $\hat{O}$, its parity (p) projection $\hat{O}_p$ is defined via

$$\langle x'|\hat{O}_p|x\rangle \equiv \langle Y_p^{x'}|\hat{O}|Y_p^x\rangle = \langle x'|\hat{\Pi}_p\hat{O}\hat{\Pi}_p|x\rangle. \tag{5.4}$$

That is, $\hat{O}_p \equiv \hat{\Pi}_p\hat{O}\hat{\Pi}_p$ in a coordinate-free representation. Therefore, the interaction V can be split into even- and odd-wave potentials, as

$$V = V_+ + V_- = \hat{\Pi}_+V\hat{\Pi}_+ + \hat{\Pi}_-V\hat{\Pi}_-, \tag{5.5}$$

which, in the momentum representation take the form

$$\langle k'|V_p|k\rangle = \frac{1}{2}[V(k - k') + pV(k + k')], \tag{5.6}$$

where $V(q)$ is the usual Fourier transform of the local potential,

$$V(q) = \int dz\, V(z)e^{iqz}. \tag{5.7}$$

Of course, the scattering operator $\hat{S}$ can also be split into even and odd waves. We can write

$$\hat{S} = \hat{S}_+ + \hat{S}_-, \tag{5.8}$$

where obviously $\hat{S}_p = \hat{\Pi}_p\hat{S}\hat{\Pi}_p$. Writing the scattering operator in the usual way,

$$\hat{S} = 1 - i\hat{T}^{\text{on}}, \tag{5.9}$$

with $\hat{T}^{\text{on}}$ the on-shell T-matrix, defined via

$$\langle k'|\hat{T}^{\text{on}}|k\rangle = 2\pi\delta(\epsilon(k) - \epsilon(k'))T^{\text{on}}(k) \tag{5.10}$$

we have $\hat{S}_p = 1_p - i\hat{T}_p^{\text{on}}$, and $\hat{T}_p$ (on- and off-shell) satisfies the partial wave Lippmann–Schwinger equation

$$\hat{T}_p(z) = V_p + V_pG_{0,p}(z)\hat{T}_p(z). \tag{5.11}$$

The partial wave scattering operator is unitary in the respective parity sector, that is,

$$\hat{S}_p^\dagger \hat{S}_p = 1_p, \tag{5.12}$$

which of course is consistent with the unitarity of the full scattering operator, recalling that $1 = 1_+ + 1_-$. Equation (5.12) leads to the partial wave optical theorem

$$\left[\hat{T}_p^{\text{on}}(z)\right]^\dagger \left[\hat{T}_p^{\text{on}}(z)\right] = i\left\{\left[\hat{T}_p^{\text{on}}(z)\right]^\dagger - \left[\hat{T}_p^{\text{on}}(z)\right]\right\}. \tag{5.13}$$

We can also get the definition of phase shifts and the form of the on-shell T-matrix in the momentum representation from equation (5.12). Since for even and odd waves only positive momentum states are necessary—the negative momentum states are automatically attached by the $\hat{\Pi}_p$ operator—we have

$$\langle k'|\hat{S}_p^\dagger \hat{S}_p|k\rangle = \int_0^\infty \frac{dq}{2\pi}\langle k'|\hat{S}_p^\dagger|q\rangle\langle q|\hat{S}_p|k\rangle = \frac{1}{2}[2\pi\delta(k-k')]. \tag{5.14}$$

Using now

$$\langle k'|\hat{S}_p|k\rangle \equiv 2\pi\delta(\epsilon(k) - \epsilon(k'))s_p(k), \tag{5.15}$$

equation (5.14) becomes

$$\langle k'|\hat{S}_p^\dagger \hat{S}_p|k\rangle = \int_0^\infty \frac{dq}{2\pi}\frac{2\pi\delta(q-k')2\pi\delta(q-k)}{\frac{2\hbar^2 k'}{m}\frac{2\hbar^2 k}{m}}|s_p(k)|^2 = \frac{1}{2}[2\pi\delta(k-k')], \tag{5.16}$$

which implies that

$$s_p(k) = \frac{\hbar^2 k}{m}e^{2i\delta_p(k)}, \tag{5.17}$$

where $\delta_p(k)$ are real numbers, called the partial wave phase shifts.

We are finally in a position to relate the on-shell T-matrix with the partial wave phase shifts. Equation (5.9) for partial waves and in the momentum representation reads (k', $k > 0$)

$$\langle k'|\hat{S}_p|k\rangle = 2\pi\delta(\epsilon(k) - \epsilon(k'))\left[\frac{\hbar^2 k}{m} - iT_p^{\text{on}}(k)\right]. \tag{5.18}$$

Using equations (5.15) and (5.17), we easily identify

$$\text{Re}\left(\frac{1}{T_p^{\text{on}}(k)}\right) = -\frac{m}{2\hbar^2 k}\frac{1}{\cot\delta_p(k)}, \tag{5.19}$$

$$\text{Im}\left(\frac{1}{T_p^{\text{on}}(k)}\right) = \frac{m}{2\hbar^2 k}, \tag{5.20}$$

so the final expression for the on-shell partial wave T-matrix takes the form

$$T_p^{\text{on}}(k) = \frac{2\hbar^2}{m}\frac{1}{-\dfrac{\tan\delta_p}{k} + i/k}. \tag{5.21}$$

It is convenient to rewrite the odd-wave ($p = -1$) amplitudes above by defining $\theta_-(k) = -\delta_-(k) + \pi/2$, and multiply numerator and denominator in equation (5.21) by k^2, so that

$$T_-^{\text{on}}(k) = \frac{2\hbar^2}{m}\frac{k^2}{-k\cot\theta_- + ik}. \tag{5.22}$$

The reason for this is that, since odd-wave interactions apply to fermions (antisymmetric wave functions), the amplitude must vanish as $k \to 0$. It is also easy to see that it behaves as $\sim k^2$ for $k \to 0$. To do this, take equation (5.6) for odd waves, and consider a smooth two-body interaction $V(q)$, which then only depends on q^2 (it is symmetric $V(q) = V(-q)$). Expand the arguments in equation (5.6), as $q^2 = k^2 + k'^2 - 2k'k$, and $\tilde{q} \equiv (k + k')^2 = k^2 + k'^2 + 2kk'$. The odd-wave potential can be formally expanded in series as

$$\begin{aligned} V_p(k', k) &= \frac{1}{2}\sum_{n=1}^{\infty} v_-^{(n)}[(k^2 + k'^2 - 2k'k)^n - (k^2 + k'^2 + 2k'k)^n] \\ &= \alpha(k'^2, k^2)k'k, \end{aligned} \tag{5.23}$$

where α is a symmetric function of k'^2 and k^2 only, and $\alpha(0, 0) = -2v_-^{(1)}$ is a non-zero constant in general. From the Lippmann–Schwinger equation for the T-matrix, it immediately follows that for small k', k, we have $\langle k'|\hat{T}_-(z)|k\rangle \propto k'k$, and therefore $T_-^{\text{on}}(k) \propto k^2$. Therefore, we may conclude that $\lim\limits_{k\to 0} k\cot\theta_-(k)$ is finite. An analogous treatment of the even-wave T-matrix leads us to concluding that $\text{Re}(1/T^{\text{on}}(k)) \propto 1$ for small k, and therefore, $\lim\limits_{k\to 0}\tan\delta_+(k)/k$ is finite. These limits specify the even- and odd-wave scattering lengths, a_e and a_o (which we shall call a indistinctively when there is no room for confusion). The odd-wave scattering length is simply

$$\frac{1}{a_o} = -\lim_{k\to 0} k\cot\theta_-, \tag{5.24}$$

while the even-wave scattering length is defined via

$$a_e = \lim_{k\to 0}\frac{\tan\delta_+}{k}. \tag{5.25}$$

We now continue by studying collision theory in the position representation. With the physical boundary conditions, the Lippmann–Schwinger equation for a scattering wave function with energy $E = \hbar^2k^2/m$ reads

$$|\psi\rangle = |\psi^{(0)}\rangle + G_0(E + i0^+)V|\psi\rangle, \tag{5.26}$$

where $|\psi^{(0)}\rangle$ is the non-interacting incident state. To write down the partial wave equivalent of equation (5.26) above, we note that $[\hat{\Pi}_p, H] = [\hat{\Pi}_p, G_0] = 0$, and freely use the fact that $\hat{\Pi}_p^n = \hat{\Pi}_p$ and that $\hat{\Pi}_p\hat{\Pi}_{-p} = 0$. Therefore, we can write

$$|\psi_p\rangle = |\psi_p^{(0)}\rangle + G_{0,p}(E + i0^+)V|\psi_p\rangle. \tag{5.27}$$

In the position representation, the partial wave Green's function takes the form $(x, x' > 0)$

$$\langle x|G_{0,p}(E + i0^+)|x'\rangle = \frac{1}{2}\left[G_0(E + i0^+;x - x') + pG_0(E + i0^+;x + x')\right], \tag{5.28}$$

where we have

$$G_0(E + i0^+;s) = -i\frac{m}{2\hbar^2 k}e^{ik|s|}. \tag{5.29}$$

Since for partial waves we restrict ourselves to $x, x' > 0$, the partial wave non-interacting Green's function (5.28) takes the asymptotic form

$$\langle x|G_{0,p}(E + i0^+)|x'\rangle \to -i\frac{m}{2\hbar^2 k}e^{ikx}\left[\frac{e^{-ikx'} + pe^{ikx'}}{2}\right], \quad x \to \infty. \tag{5.30}$$

Inserting equation (5.30) into the position representation of equation (5.27), we obtain, as $x \to \infty$,

$$\psi_+(x) \to \cos(kx) - \frac{im}{2\hbar^2 k}e^{ikx}\int_0^\infty dx' \cos(kx')V(x')\psi_+(x'), \tag{5.31}$$

$$\psi_-(x) \to i\sin(kx) + \frac{m}{2\hbar^2 k}e^{ikx}\int_0^\infty dx' \sin(kx')V(x')\psi_-(x'). \tag{5.32}$$

We identify now

$$\int_0^\infty dx' \cos(kx')V(x')\psi_+(x') = \langle k|V_+|\psi_+\rangle = \langle k|\hat{T}_+(E + i0^+)|k\rangle = T_+^{\text{on}}(k), \tag{5.33}$$

$$i\int_0^\infty dx' \sin(kx')V(x')\psi_-(x') = \langle k|V_-|\psi_-\rangle = \langle k|\hat{T}_-(E + i0^+)|k\rangle = T_-^{\text{on}}(k). \tag{5.34}$$

Therefore, inserting the above relations together with equation (5.21) into the asymptotic expressions (5.31) and (5.32), we obtain, as $x \to \infty$,

$$\psi_+(x) \to ie^{-i\delta_+}\sin(kx + \delta_+(k)), \tag{5.35}$$

$$\psi_-(x) \to ie^{-i\delta_-}\cos(kx - \delta_-(k)). \tag{5.36}$$

If one wishes to extend the results for large negative x, then x is replaced with $|x|$ in equations (5.35) and (5.36), and the right-hand-side of equation (5.36) must be multiplied by sgn(x). In equation (5.36), we may write the phase shift $\theta_- = -\delta_- + \pi/2$, as in equation (5.22), and write the asymptotic expression as

$$\psi_-(x) \rightarrow e^{-i\theta_-}\mathrm{sgn}(x)\sin(k|x| + \theta_-(k)), \ \ |x| \rightarrow \infty. \qquad (5.37)$$

5.2 Dimensional reduction from three to one dimension

We move on now to the core of this chapter, namely the dimensional reduction of the three-dimensional two-body scattering problem in a two-dimensional transversal trap. We do this for two bosons, or equivalently two singlet fermions. The case of triplet or spinless fermions is more complicated, and we refer the reader to reference [2], as well as the regularization–renormalization of the odd-wave interaction in references [3, 4].

5.2.1 Two bosons in a harmonic trap: Olshanii's method

We consider two bosonic particles that interact via leading-order EFT in three spatial dimensions. In order for these to behave as if kinematically constrained to one spatial dimension, we require an anisotropic trapping potential that is much tighter in two directions, say x and y, than in the other direction z. For simplicity, we take the external potential V to act only in the transversal direction $\mathbf{r}_\perp = (x, y)$. The two-body Hamiltonian takes the form

$$H = \frac{\mathbf{p}_1^2}{2m} + \frac{\mathbf{p}_2^2}{2m} + V(\mathbf{r}_{\perp,1}) + V(\mathbf{r}_{\perp,2}) + V_{\mathrm{LO}}(\mathbf{r}_1 - \mathbf{r}_2). \qquad (5.38)$$

Among all possible equivalent choices of regularization–renormalization for V_{LO}, we shall choose the Lee–Huang–Yang pseudopotential (see section 6.1.2), which is well suited to the problem at hand. That is, we choose

$$V_{\mathrm{LO}}(\mathbf{r}) = g_{\mathrm{R}}\delta(\mathbf{r})\partial_r(r\ \cdot\ ), \qquad (5.39)$$

where $\mathbf{r} = \mathbf{r}_1 - \mathbf{r}_2$, $r = |\mathbf{r}|$, and $g_{\mathrm{R}} = 4\pi\hbar^2 a/m$ is the renormalized LO coupling constant, with a the three-dimensional scattering length.

For a general external trapping potential, equation (5.38) is not separable into centre-of-mass and relative coordinates in the transversal direction. The case of a harmonic trap, $V \propto r_\perp^2$, is the simplest one since it allows for this separation. Moreover, it is quite relevant for two reasons. Firstly, harmonic traps are available in experiments with ultracold atoms and, secondly, most trapping potentials (such as optical lattices) are quadratic around their minima, and a harmonic approximation is reasonable. The potential is given by

$$V(\mathbf{r}_{\perp,i}) = \frac{1}{2}m\omega_\perp^2(x_i^2 + y_i^2). \qquad (5.40)$$

After separation of centre-of-mass and relative coordinates, the relative Hamiltonian takes the form

$$H = \frac{\mathbf{p}^2}{m} + \frac{1}{2}V(\mathbf{r}_\perp) + V_{\mathrm{LO}}(\mathbf{r}). \qquad (5.41)$$

Above, $\mathbf{p}$ is the relative momentum. The eigenstates of the free transverse Hamiltonian $H_\perp = \mathbf{p}_\perp^2/m + (1/2)V_{\mathbf{r}_\perp}$ are harmonic oscillator eigenstates of the form

$$\Phi_{n\ell}(r_\perp, \phi) = u_n(r_\perp)\frac{e^{i\ell\phi}}{\sqrt{2\pi}}, \tag{5.42}$$

where $n = 0, 1, \ldots,$ and $\ell = -n, -n+2, \ldots, n$ and associated eigenenergies $E_{n\ell} = \hbar\omega_\perp(n+1)$. For our purposes, the concrete shape of the radial functions u_n in equation (5.42) is inconsequential. Since the two-body interaction has a zero-range, only solutions $\Phi_{n\ell}$ that are non-zero at $r_\perp = 0$, and their values at $r_\perp = 0$ are relevant. Therefore, we only need to know that $\Phi_{n\ell}(0, \phi) = 0$ for $\ell \neq 0$, and that $|\Phi_{n,0}(0, \phi)|^2 = 1/\pi a_\perp^2$, where $a_\perp = \sqrt{2\hbar/m\omega_\perp}$ is the oscillator length.

The one-dimensional scattering solutions in the transversally trapped ground state Φ_{00} only exist if their energies lie below the first excited state (accessible via zero-range s-wave collisions, that is, $n = 2$) in the trap. If k is the incident momentum in the z-direction, this means

$$\frac{\hbar^2 k_z^2}{m} + \hbar\omega_\perp < 3\hbar\omega_\perp \Rightarrow \frac{\hbar^2 k_z^2}{m} < 2\hbar\omega_\perp. \tag{5.43}$$

This scattering state ψ must have, asymptotically, the form

$$\psi(r_\perp, \phi, z) \to \sin(k|z| + \theta_k)\Phi_{00}(r_\perp, \phi) \equiv \psi_{1D}(z)\Phi_{00}(\mathbf{r}_\perp), \quad |z| \to \infty. \tag{5.44}$$

Note that, since the wave function above is three-dimensional, the asymptotic limit is strict, in contrast with the 'genuine' one-dimensional, zero-range case, where the asymptotic form is correct for all z. Hence, we can write the state ψ as

$$\psi(\mathbf{r}) = \psi_{1D}(z)\Phi_{00}(\mathbf{r}_\perp) + F(\mathbf{r}), \tag{5.45}$$

where $F(\mathbf{r})$ is perpendicular to Φ_{00}, that is,

$$\int d\mathbf{r}_\perp \Phi_{00}^*(\mathbf{r}_\perp)F(\mathbf{r}) = 0. \tag{5.46}$$

With all the above discussion, we perform the following expansion for F,

$$F(\mathbf{r}) = \sum_{n=1}^{\infty} f_{2n} e^{-\lambda_{2n}|z|}\Phi_{2n,0}(\mathbf{r}_\perp). \tag{5.47}$$

Inserting equation (5.45) into the stationary Schrödinger equation $H\psi = E\psi$, with $E = \hbar^2 k^2/m + \hbar\omega_\perp$, and using the expansion in equation (5.47), we have,

$$\hbar\omega_\perp(2n+1) - \frac{\hbar^2\lambda_{2n}^2}{m} = E, \tag{5.48}$$

and

$$\left[\frac{2\hbar^2 k}{m}\cos\theta_k \Phi_{00}(\mathbf{r}_\perp) + \frac{2\hbar^2}{m}\sum_{n=1}^{\infty} f_{2n}\Phi_{2n}(\mathbf{r}_\perp)\lambda_{2n}\right]\delta(z) + g_{\mathrm{R}}\frac{\partial}{\partial r}[r\psi]\delta(\mathbf{r}) = 0. \tag{5.49}$$

We now define

$$\eta = \lim_{\mathbf{r}\to 0}\frac{\partial}{\partial r}[r\psi(\mathbf{r})] = \lim_{|z|\to 0}\frac{\partial}{\partial |z|}[|z|\psi(\mathbf{r}_\perp = 0, z)]. \tag{5.50}$$

Multiplying equation (5.49) by $\Phi_{00}^*(\mathbf{r}_\perp)$ and integrating over $\mathbf{r}_\perp$, we obtain

$$\eta = -\frac{2\hbar^2\sqrt{\pi}a_\perp}{mg_\mathrm{R}}k\cos\theta_k, \tag{5.51}$$

while doing the same but with $\Phi_{2n,0}^*$, we obtain

$$f_{2n} = -\frac{\eta m g_\mathrm{R}}{2\hbar^2\sqrt{\pi}a_\perp}\frac{1}{\lambda_{2n}}. \tag{5.52}$$

We are going to make use of the Euler–Maclaurin summation formula, which will allow us to compute η. Choosing $n_c > 1$, the following approximation holds for the sum of a function h_n with a smooth extension to the reals (which we also call h_n) and with $\lim_{n\to\infty} h_n = 0$,

$$\sum_{n=1}^{\infty} h_n \approx \sum_{n=1}^{n_c} h_n - \frac{h_{n_c}}{2} + \int_{n_c}^{\infty} dn h_n. \tag{5.53}$$

Since for η we can write, explicitly,

$$\eta = \frac{1}{\sqrt{\pi}a_\perp}\sin\theta_k + \frac{1}{\sqrt{\pi}a_\perp}\lim_{|z|\to 0}\partial_{|z|}\left[|z|\sum_{n=1}^{\infty} f_{2n}e^{-\lambda_{2n}|z|}\right], \tag{5.54}$$

we will approximate the following sum

$$s(z) = \sum_{n=1}^{\infty}\frac{e^{-\lambda_{2n}|z|}}{\lambda_{2n}}, \tag{5.55}$$

using equation (5.53), as

$$s(z) \approx \sum_{n=1}^{n_c}\frac{e^{-\lambda_{2n}|z|}}{\lambda_{2n}} - \frac{e^{-\lambda_{2n_c}|z|}}{2\lambda_{2n_c}} + \frac{e^{-\lambda_{2n_c}|z|}}{|z|} \equiv s_0(z) + \frac{e^{-\lambda_{2n_c}|z|}}{|z|}. \tag{5.56}$$

Above, we can clearly see the $1/|z|$ divergence as $|z| \to 0$ that we need to eliminate via the pseudopotential. Therefore, we obtain the following (n_c-dependent) approximation for the limit in equation (5.54),

$$\lim_{|z|\to 0}\partial_{|z|}\left[|z|\sum_{n=1}^{\infty} f_{2n}e^{-\lambda_{2n}|z|}\right] \approx -\frac{\eta m g_\mathrm{R}}{2\hbar^2\sqrt{\pi}a_\perp}[s_0(0) - \lambda_{2n_c}]. \tag{5.57}$$

In fact, it is not difficult to see that equation (5.57) is exact and finite as $n_c \to \infty$, and therefore we can write

$$\lim_{|z|\to 0}\partial_{|z|}\left[|z|\sum_{n=1}^{\infty}f_{2n}e^{-\lambda_{2n}|z|}\right] = -\frac{\eta m g_{\rm R}}{2\hbar^2\sqrt{\pi}a_\perp}\lim_{n_c\to\infty}[s_0(0) - \lambda_{2n_c}] \equiv -\frac{\eta m g_{\rm R}}{2\hbar^2\sqrt{\pi}a_\perp}L_k. \quad (5.58)$$

Putting this all together, we finally obtain

$$[-k\cot\theta_k]^{-1} = \frac{2\hbar^2}{m g_{\rm R}}\pi a_\perp^2\left[1 + \frac{m g_R}{2\hbar^2}\frac{L_k}{\sqrt{\pi}a_\perp}\right]. \quad (5.59)$$

We may simplify the above expression by identifying the renormalized coupling constant $g_R = 4\pi\hbar^2 a/m$, where a is the three-dimensional scattering length, obtaining

$$-k\cot\theta_k = \frac{a}{a_\perp^2}\frac{1}{1 + (2\sqrt{\pi}/a_\perp)L_k}. \quad (5.60)$$

The reader is encouraged to extract the one-dimensional scattering length from equation (5.60) (see the problem section). The main feature of equation (5.60) is the existence of a so-called 'confinement-induced resonance' (CIR), which occurs when $-1/k\cot\theta_k$ vanishes, that is, for

$$a_\perp = -\frac{2\sqrt{\pi}}{L_k}. \quad (5.61)$$

5.2.2 Two bosons in a harmonic trap: Lüscher's method

The method used in the previous subsection is essentially restricted to that (and some slightly more general) particular case. Another method, which requires more numerical work, but is easily generalizable, can be devised by using diagonalization in finite boxes, with the ideas developed in chapter 2.

For simplicity, we will separate centre-of-mass and relative coordinates prior to performing any calculations. The two-body interaction in this case is fully general, aside from short ranged, not necessarily restricted to LO EFT. We place the reduced system in a large box with sides of lengths (L_x, L_y, L_z). The boundary conditions in the x- and y-directions do not necessarily coincide with that in the z-direction, and $L_\perp \equiv L_x = L_y \neq L_z$ in general. For simplicity, we will use open boundary conditions, such that $\psi(\pm L_\perp/2, y, z) = \psi(x, \pm L_\perp/2, z) = \psi(x, y, \pm L_z/2) = 0$. We shall choose $L_\perp$ large enough that the low-lying eigenenergies of $H_\perp$ are sufficiently accurate, and vary L_z to obtain a nice fit to the energy spectrum of the target quasi-one-dimensional collisions.

In a pure one-dimensional collision, the scattering states in the relative coordinate z take the simple form

$$\psi_k(z) = \sin(k|z| + \theta_k),\ |z| > R, \quad (5.62)$$

with R the range of the interaction. Open boundary conditions in one dimension allow for an extremely simple quantization condition, namely $\psi_k(\pm L_z/2) = 0$ implies

$$\frac{kL_z}{2} + \theta_k = n\pi,\ n = 1, 2, \ldots, \quad (5.63)$$

and the energy spectrum is given by $\epsilon(k) = \hbar^2 k^2/m$. Therefore, the phase shifts are very easily calculated.

If equation (5.62) remained valid strictly for $|z| > R$ in the fully three-dimensional problem with an external potential in the transverse direction, then the low-lying energy spectrum would simply be $E_\perp + \hbar^2 k^2/m$, with $E_\perp$ the transverse ground state energy, with the same quantization rule. This is not entirely the case, and accurate determination of the phase shifts could require extra work. Let us examine what happens in the LO EFT case. The scattering states are given by equation (5.45). The function F in equation (5.47) is non-trivial asymptotically. We may set $\mathbf{r}_\perp = 0$, and define $f(z) \equiv F(\mathbf{r}_\perp = 0, z)$. The function f can be obtained from equations (5.47), (5.48) and (5.52). At low energy ($k \to 0$), we can estimate the asymptotic form of f as

$$f(z) \propto \sum_{n=1}^{\infty} \frac{e^{-c\sqrt{n}|z|}}{\sqrt{n}}, \tag{5.64}$$

with $c > 0$ a constant. For large z, we may estimate f, via the Euler–Maclaurin formula, by changing the sum in equation (5.64) for an integral over n. This gives us $f(z)$ tending to zero at large z faster than an exponential. Therefore, we may safely (up to exponentially small corrections) use the standard one-dimensional quantization to extract phase shifts from exact diagonalization.

To make computations as easy as possible, we will discretize the Hamiltonian on a lattice with spacing d and also use LO EFT on the lattice. The simplest bosonic L-EFT has the following Hamiltonian

$$H = H_0 + U\delta_{\mathbf{r},0}. \tag{5.65}$$

The relation between the Hamiltonian (5.65) and the scattering length will be found in section 6.1.5. There, this is done for the one-body problem, which means that here we must take the mass in section 6.1.5 and divide it by two, that is, replace m with $\mu = m/2$. The particular expression we are interested in is equation (6.76). Although we have previously stated that the scattering phase shifts can be extracted with ease via exact diagonalization, there may be two-body bound states when the system is dimensionally reduced even if the three-dimensional system has no bound state. This is the case, for instance, on resonance. When this is the case, applying periodic boundary conditions in the z-direction (the free dimension) and open boundary conditions in the trapped direction, we expect a size dependence of the binding energy as follows

$$E_{\rm B}(L) \approx E_{\rm B,\infty} + \alpha \exp(-\beta L_z). \tag{5.66}$$

Above, α and β are constants that the reader is encouraged to find, and $E_{B,\infty}$ is the actual binding energy for $L_z \to \infty$. On resonance, we estimate $\epsilon_{\rm B,\infty} \equiv E_{\rm B,\infty}/\hbar\omega_\perp$ using diagonalization of the Hamiltonian with a finite lattice spacing, and a fit to the binding energy of the form of equation (5.66) to be (see figure 5.1)

$$\epsilon_{\rm B,\infty} = 0.6204 \pm 10^{-4}. \tag{5.67}$$

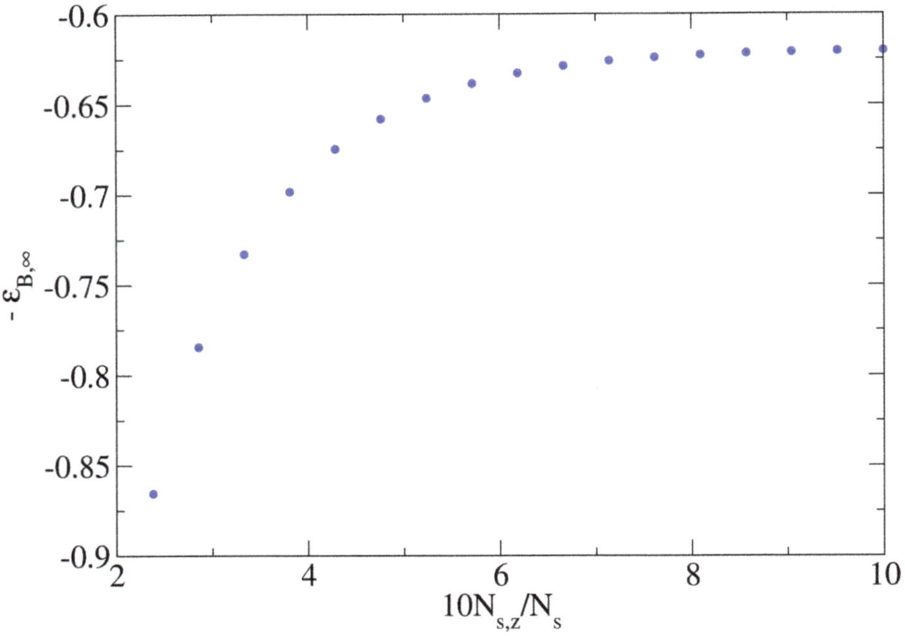

Figure 5.1. (Minus) the binding energy in dimensionless form (see text) as a function of the size of the one-dimensional channel, with fixed $L_\perp$ and $N_s = 21$, where N_s is the number of lattice sites in each of the trapped directions.

This value is to be compared with the exact expression, which can be obtained analytically as well (see reference [5]), and gives $E_{B,\infty} = 0.60\ldots\hbar\omega_\perp$. The discrepancy is due to the finite resolution of the lattice, and can be improved systematically by obtaining this for different lattice spacings and extrapolating the result to the continuum limit.

Problems

1. Prove that equation (5.66) holds, and calculate the constants α and β.
2. Calculate the one-dimensional scattering length and effective range for a two-boson system under two-dimensional harmonic trapping by taking the low-energy limit in equation (5.60).
3. Obtain the one-dimensional scattering length, with the same hypotheses of the previous problem, using, instead, an ansatz for $\psi_{1D}(z) = |z| - a$ at zero collision energy.

References

[1] Barlette V E, Leite M M and Adhikari S K 2000 Quantum scattering in one dimension *Eur. J. Phys.* **21** 435

[2] Granger B E and Blume D 2004 Tuning the interactions of spin-polarized fermions using quasi-one-dimensional confinement *Phys. Rev. Lett.* **92** 133202

[3] Cui X 2016 universal one-dimensional atomic gases near odd-wave resonance *Phys. Rev.* **A94** 043636

[4] Valiente M and Zinner N T 2015 Effective field theory of interactions on the lattice *Few-Body Syst.* **56** 845

[5] Bergeman T, Moore M G and Olshanii M 2003 Atom-atom scattering under cylindrical harmonic confinement: numerical and analytic studies of the confinement induced resonance *Phys. Rev. Lett.* **91** 163201

IOP Publishing

Strongly Interacting Quantum Systems, Volume 1

Few-body physics

Manuel Valiente and Nikolaj Thomas Zinner

Chapter 6

One- and two-body physics

In this chapter, we take a deeper look at a number of interesting and important models that contain one or two particles. We focus on examples, from a variety of settings and fields, with the goal of not only finding relevant physical phenomena occurring in these systems, but also giving the reader a place where they can actually practise much of what has been said previously throughout the book. In some cases, straight application of the methods explained earlier suffices, while in some others slightly different (yet related) methods are needed. Most of the models studied in this chapter are justified in chapters 4 and 5, sometimes in full. Other times, the justification from chapters 4 and 5 may be only partial, and the interested reader is encouraged to consult more specialized literature in the subject.

6.1 One-body problems

We begin with the simpler, single-particle systems. These usually have an infinitely heavy impurity of sorts, that is, a static potential located somewhere in the system, while what we are interested in is the 'active', itinerant particle. For non-relativistic, homogeneous problems, for which centre-of-mass and relative coordinates separate, we study here the particle-impurity scattering, and avoid having to study the two-body problem later on, since one derives from the other. As we stated earlier, we focus on examples, beginning with the one-dimensional Dirac delta interaction.

6.1.1 Dirac delta interaction in one dimension

The LO EFT interaction for an impurity in one dimension is given, as usual, by a constant in the momentum representation. The one-body problem is regular, and no regularization is required. Renormalization, on the other hand, is actually needed in order to have the correct low-energy scattering properties, in this case the (even-wave) scattering length. The LO interaction, in the position representation, is local,

doi:10.1088/978-0-7503-3087-9ch6

and given by a Dirac delta potential. That is, the Hamiltonian of the particle is given by

$$H = \frac{p^2}{2m} + g\delta(z). \tag{6.1}$$

Since the interaction has a zero range, the even-wave stationary scattering states with energy $E = \hbar^2k^2/2m$ must have the following form everywhere (see chapter 5),

$$\psi(z) = \sin(k|z| + \theta_k). \tag{6.2}$$

Above, θ_k is the even-wave phase shift. Introducing equation (6.2) into the Schrödinger equation, we obtain

$$\tan\theta_k = \frac{\hbar^2 k}{mg}. \tag{6.3}$$

To obtain the above relation, we have used the following distributional derivatives,

$$\frac{d}{dx}|x| = \mathrm{sgn}(x), \tag{6.4}$$

$$\frac{d}{dx}\,\mathrm{sgn}(x) = 2\delta(x). \tag{6.5}$$

From equation (6.3), we can obtain the scattering length as

$$a = -\lim_{k\to 0}\frac{\tan\theta_k}{k} = -\frac{\hbar^2}{mg}. \tag{6.6}$$

Regarding odd-wave scattering, since the Dirac delta interaction does not act on antisymmetric states, there is no scattering. That is, the odd-wave stationary scattering states are given by

$$\psi(x) = \sin(kx). \tag{6.7}$$

Let us take a look now at bound states, which may exist if $g < 0$. Again, due to the zero range of the Dirac delta interaction, the even (unnormalized) bound states must take the following form everywhere

$$\psi(x) = e^{-\lambda|x|}, \tag{6.8}$$

with $\lambda > 0$. Inserting equation (6.8) into the stationary Schrödinger equation, and using the identities (6.4) and (6.5), we obtain, for the bound state energy

$$E = -\frac{mg^2}{2\hbar^2}, \tag{6.9}$$

and $\lambda = -mg/\hbar^2$.

This is essentially all we need to say about the one-body (and the two-body) problem at low energies in one dimension in the even-wave channel. The odd-wave channel is slightly more involved, as it requires regularization–renormalization, and we leave this as a problem for the reader in the end of the chapter.

6.1.2 Dirac delta interaction in three dimensions

The problem of a one non-relativistic particle in three dimensions interacting with a (regularized–renormalized) Dirac delta impurity has been solved already, and in fact more involved problems with it have been solved already. However, we can still solve it once more in a direct manner using the Huang–Yang pseudopotential. Hence, we start by defining the renormalized coupling constant

$$g_R = \frac{2\pi\hbar^2 a}{m}, \tag{6.10}$$

where a is the s-wave scattering length. The Hamiltonian is given by

$$H = -\frac{\hbar^2}{2m}\nabla^2 + g_R\delta(\mathbf{r})\frac{\partial}{\partial r}(r\ \cdot). \tag{6.11}$$

All partial-wave scattering states of the stationary Schrödinger equation are non-interacting states except for the s-wave state. We solve the stationary Schrödinger equation at zero energy. To do that, we take the zero energy state

$$\psi(\mathbf{r}) = \frac{1}{r} - \frac{1}{a}, \tag{6.12}$$

and use the fact that

$$-\nabla^2\frac{1}{r} = 4\pi\delta(\mathbf{r}). \tag{6.13}$$

We immediately see that ψ is a solution to the Schrödinger equation at zero energy.

6.1.3 Short-range impurity in graphene: Klein tunneling

This is the first somewhat non-trivial example we explore in this chapter. We may imagine a perfect sheet of graphene, and the electrons near the Fermi points being essentially non-interacting. Then, a non-magnetic defect (whether controllable or otherwise), that interacts with the electrons via a short-range potential, is placed. We wonder what the scattering properties of the electrons off the impurity are, at low energies.

To study this problem, we use directly the continuum, effective, low-energy Hamiltonian of a free electron in graphene. The non-interacting Hamiltonian, in the first quantization and in the space representation, is given by

$$H_0 = -iv_F\vec{\sigma}\cdot\nabla, \tag{6.14}$$

where $\vec{\sigma} = (\sigma_x, \sigma_y)$ is the two-dimensional vector of spin-1/2 Pauli matrices, ∇ is the two-dimensional gradient, and v_F is the Fermi velocity divided by $\hbar$ (equivalently, we may set $\hbar = 1$ and say v_F is the Fermi velocity). The eigenstates of H_0 in equation

(6.14) are spinors, $\psi^\dagger(x) = (\psi_a^*, \psi_b^*)$, with the subindices a and b referring, in the original tight-binding model of graphene, to sublattices a and b. Since the non-interacting system is homogeneous, the eigenstates can be written as

$$\psi(x) = e^{i\mathbf{k}\cdot\mathbf{r}}\phi, \tag{6.15}$$

where ϕ is a constant spinor. The stationary Schrödinger equation reduces to

$$v_F\left(k_x - ik_y\right)\phi_b = E\phi_a, \tag{6.16}$$

$$v_F\left(k_x + ik_y\right)\phi_a = E\phi_b, \tag{6.17}$$

from where it follows that there are two (touching) bands, that we label $\lambda = \pm$, with energies $E_\pm(\mathbf{k}) = \pm v_F|\mathbf{k}|$, and eigenstates

$$\phi_{a,\,\pm}^{\mathbf{k}} = e^{-i\theta_\mathbf{k}/2}/\sqrt{2}, \tag{6.18}$$

$$\phi_{b,\,\pm}^{\mathbf{k}} = \pm e^{i\theta_\mathbf{k}/2}/\sqrt{2}, \tag{6.19}$$

where $\theta_\mathbf{k}$ is the polar angle in two-dimensional momentum space.

Following what we have done many times so far, we try and simplify our calculations by considering an LO impurity interaction potential. That is, defining a bare coupling constant g_0, we take

$$\langle\mathbf{k}'|V|\mathbf{k}\rangle = g_0. \tag{6.20}$$

Since in graphene there are two (touching) bands, we shall call the non-interacting eigenstates of H_0 in ket notation $|\mathbf{k}\lambda\rangle$, with $\lambda = \pm$. The matrix elements of the interaction become slightly less trivial, namely

$$\langle\mathbf{k}'\lambda'|V|\mathbf{k}\lambda\rangle = g_0\cos\left[\frac{\theta_{\mathbf{k}'} - \theta_\mathbf{k}}{2}\right], \quad \lambda = \lambda', \tag{6.21}$$

$$= ig_0\sin\left[\frac{\theta_{\mathbf{k}'} - \theta_\mathbf{k}}{2}\right], \quad \lambda \neq \lambda'. \tag{6.22}$$

The above matrix elements can also be written in a more compact form, as

$$\langle\mathbf{k}'\lambda'|V|\mathbf{k}\lambda\rangle = \frac{g_0}{2}\left[f_{\mathbf{k}'}f_\mathbf{k}^* + \lambda\lambda' f_{\mathbf{k}'}^* f_\mathbf{k}\right], \tag{6.23}$$

where we have defined

$$f_\mathbf{k} = e^{i\theta_\mathbf{k}/2}. \tag{6.24}$$

We are ready now to write down and solve the Lippmann–Schwinger equation for the T-matrix. In principle, the T-matrix takes the following general form

$$\langle\mathbf{k}'\lambda'|T(z)|\mathbf{k}\lambda\rangle = A_{\lambda\lambda'}f_{\mathbf{k}'}f_\mathbf{k}^* + B_{\lambda\lambda'}f_{\mathbf{k}'}^*f_\mathbf{k} + C_{\lambda\lambda'}f_{\mathbf{k}'}f_\mathbf{k} + D_{\lambda\lambda'}f_{\mathbf{k}'}^*f_\mathbf{k}^*. \tag{6.25}$$

Above, A, B, C and D depend on z. However, it is easy to see that $C_{\lambda\lambda'} = D_{\lambda\lambda'} = 0$, and the Lippmann–Schwinger equation reduces to the following set of two decoupled equations for $A_{\lambda\lambda'}$ and $B_{\lambda\lambda'}$,

$$A_{\lambda\lambda'} = \frac{g_0}{2} + \frac{g_0}{2}\sum_{\gamma=\pm}\int\frac{d^2q}{(2\pi)^2}G_0(\gamma,\,\mathbf{q};\,z)A_{\lambda\gamma}|f_{\mathbf{q}}|^2, \tag{6.26}$$

$$B_{\lambda\lambda'} = \frac{g_0}{2}\lambda\lambda' + \frac{g_0}{2}\sum_{\gamma=\pm}\int\frac{d^2q}{(2\pi)^2}G_0(\gamma,\,\mathbf{q};\,z)B_{\lambda\gamma}\gamma\lambda'|f_{\mathbf{q}}|^2, \tag{6.27}$$

where $G_0(\gamma,\,\mathbf{q};\,z)$ is the non-interacting Green's function, given by

$$G_0(\gamma,\,\mathbf{q};\,z) = \frac{1}{z - \gamma v_F|\mathbf{q}|}. \tag{6.28}$$

The only integral we need to compute, looking at equations (6.26), (6.27) and (6.28), as well as equation (6.24), is

$$I_\gamma(z) = \int\frac{d^2q}{(2\pi)^2}\frac{1}{z - \gamma v_F|\mathbf{q}|}. \tag{6.29}$$

Clearly, I_γ is UV-divergent, and therefore we calculate it with a UV-cutoff Λ, that is

$$I_\gamma(z) = \int_0^\Lambda\frac{dq}{2\pi}\frac{q}{z - \gamma v_F q}. \tag{6.30}$$

For concreteness, let us assume that we want to go on-shell, and we choose $z = \lambda v_F k + i\eta$, with $k > 0$. The integral in equation (6.30) becomes

$$I_\gamma(z) = -\frac{1}{2\pi v_F}\left[\gamma\Lambda + \lambda k\ln\left|\frac{\Lambda}{k}\right| - i\pi k\delta_{\lambda,\gamma}\right]. \tag{6.31}$$

The above relation might appear to point to an issue with renormalizability, given that it has two different UV-divergences (linear and logarithmic), while we only have one coupling constant. This is not correct, since in fact, having two channels, we could have chosen two coupling constants (namely g_0 for the + and – channels). Not having done so, in fact, is no problem, since the linear divergences will cancel out in the solution for the T-matrix, as we shall see. However, the problem with renormalisability is actually there because of the energy dependence in the prefactor of the logarithmic term in equation (6.31), as we shall see shortly. As an example, let us compute $A_{\lambda,\lambda'}$. It is not difficult to see, from equation (6.26), that $A_{++} = A_{--} = A_{+-} = A_{-+} \equiv A$, and this is given by

$$A = \frac{1}{2/g_0 - (I_+ + I_-)}. \tag{6.32}$$

The denominator, using equation (6.31), requires

$$I_+ + I_- = \frac{1}{2\pi v_F}\left[2k \ln \left|\frac{\Lambda}{k}\right| + i\pi k\right], \tag{6.33}$$

where, as promised, the linear divergences have cancelled each other. Similar expressions hold for $B_{\lambda\lambda'}$. In particular, we have $B_{++} = A$. With the cutoff in place, the on-shell T-matrix is given by

$$\langle \mathbf{k}' + |T(z)|\mathbf{k} + \rangle = 2A \cos\left[\frac{\theta_{\mathbf{k}} - \theta_{\mathbf{k}'}}{2}\right]. \tag{6.34}$$

The renormalization condition $A(\mu + i\eta) = g_R$ renders the on-shell T-matrix trivial (i.e., zero) as the cutoff is removed ($\Lambda \to \infty$). This is impossible to fix. In fact, it is not unphysical: this phenomenon is known as 'Klein tunneling', and it refers to perfect transmission (i.e., trivial T-matrix) for a chiral system scattering off a potential barrier. In our case, this is the LO EFT version of this problem, and it is well known that realistic potentials do not fix this. Clearly, since graphene is a lattice, it has a natural cutoff and pure Klein tunnelling will not occur. However, the LO EFT approach shows that scattering of electrons near the Fermi points off short-range impurities is highly suppressed due to the chiral nature of the system.

6.1.4 Scattering off an impurity in a hypercubic lattice

We move on now to the next non-trivial one-body problem, consisting of a non-relativistic particle in a periodic potential with hypercubic symmetry in D spatial dimensions. The non-interacting Hamiltonian is quite simply given by

$$H_0 = -\frac{\hbar^2}{2m}\nabla^2 + \sum_{i=1}^{D} V(x_i), \tag{6.35}$$

with ∇^2 the D-dimensional Laplacian, and $V(x_i + d) = V(x_i)$, for $i = 1, 2, \ldots, D$, with d the lattice spacing. We shall place an impurity at the centre of the lattice (although, in this case, the placement of the impurity does matter) for simplicity, and we assume it is a low-energy interaction. That is, the interaction potential is given by

$$H_I = g_0\delta(\mathbf{r}). \tag{6.36}$$

Above, g_0 is the bare coupling constant, of course. The non-interacting eigenstates are given by Bloch states, namely

$$\psi_{n\mathbf{k}}(\mathbf{r}) = \phi_{n\mathbf{k}}(\mathbf{r})e^{i\mathbf{k}\cdot\mathbf{r}}, \qquad \mathbf{k} \in \mathrm{BZ}. \tag{6.37}$$

The Bloch functions ϕ depend on the band index n and the quasi-momentum $\mathbf{k}$. Due to the hypercubic symmetry, we may use a vector band index $\mathbf{n} = (n_1, n_2, \ldots, n_D)$, since the Bloch functions in each direction x_i ($i = 1, \ldots, D$) factorize. The incident state will be a Bloch function with band index $\mathbf{n}$ and quasi-momentum $\mathbf{k}$, corresponding to eigenenergy $\epsilon_{\mathbf{n}}(\mathbf{k})$. The Lippmann–Schwinger equation is genuinely multichannel and, in fact, has infinitely many channels, since there are infinitely many bands. It reads, in the quasi-momentum representation,

$$\langle \mathbf{n}'\mathbf{k}'|T(z)|\mathbf{nk}\rangle = \langle \mathbf{n}'\mathbf{k}'|H_I|\mathbf{nk}\rangle + \sum_{\mathbf{m}}\int_{\mathrm{BZ}} \frac{d^D q}{(2\pi)^D}\frac{\langle \mathbf{n}'\mathbf{k}'|H_I|\mathbf{mq}\rangle}{z-\epsilon_{\mathbf{m}}(\mathbf{q})}\langle \mathbf{mq}|T(z)|\mathbf{nk}\rangle. \quad (6.38)$$

The low-energy interaction, equation (6.36) offers a major simplification for the matrix elements, as usual. Being located at the centre, we easily obtain

$$\langle \mathbf{n}'\mathbf{k}'|H_I|\mathbf{nk}\rangle = g_0\phi^*_{\mathbf{n}'\mathbf{k}'}(0)\phi_{\mathbf{nk}}(0) \equiv g_0\phi^*_{\mathbf{n}'\mathbf{k}'}\phi_{\mathbf{nk}}. \quad (6.39)$$

Above, we have removed the reference to the point (the centre) at which the Bloch functions are evaluated to avoid overloading the notation later on. Inserting equation (6.39) into the Lippmann–Schwinger equation, equation (6.38), we immediately see that

$$\langle \mathbf{n}'\mathbf{k}'|T(z)|\mathbf{nk}\rangle = t_{\mathbf{n}'\mathbf{n}}(z)\phi^*_{\mathbf{n}'\mathbf{k}'}\phi_{\mathbf{nk}}, \quad (6.40)$$

where $t_{\mathbf{n}'\mathbf{n}}$ satisfies

$$t_{\mathbf{n}'\mathbf{n}}(z) = g_0 + g_0\sum_{\mathbf{m}}\int_{\mathrm{BZ}} \frac{d^D q}{(2\pi)^D}\frac{|\phi_{\mathbf{mq}}|^2}{z-\epsilon_{\mathbf{m}}(\mathbf{q})}t_{\mathbf{mn}}(z). \quad (6.41)$$

The equation above is an infinite system of coupled equations. With $\mathbf{n}$ fixed, all $\mathbf{n}'$ couple with all others. There is a way to avoid placing a cutoff on the number of bands, at the cost of another approximation. Note that, as $|\mathbf{m}| \to \infty$, the band gaps become small and perturbation theory (for the non-interacting system) becomes accurate due to the high energy. Therefore, one may approximate $\phi_{\mathbf{mq}}$ with a plane wave and, therefore,

$$|\phi_{\mathbf{mq}}|^2 \to 1, \quad (6.42)$$

as $|\mathbf{m}| \to \infty$, as well as an almost quadratic, almost continuous dispersion. The sum of the integral at large $|\mathbf{m}|$ can then be approximated with a continuous integral (in the extended zone scheme), with the advantage that the bare coupling constant g_0 can then be written in terms of a standard UV-cutoff, and takes the standard free-space form.

In one spatial dimension things can be heavily simplified, since it is possible to calculate the non-interacting Green's function in the position representation very easily. At energy E, the Green's function is a solution to

$$[E - H_0(z)]G_0(E; z, z') = \delta(z - z'). \quad (6.43)$$

If we call the two solutions (assuming parity is a good quantum number) to the non-interacting problem at energy E $\psi_{n\pm k}$, then the non-interacting Green's function takes the form

$$G_0(E; z, z') = \theta(z - z')A_{nk}(z')\psi_{nk}(z) + \bar{\theta}(z' - z)A_{n,-k}(z')\psi_{n,-k}(z), \quad (6.44)$$

where θ and $\bar{\theta}$ are Heaviside distributions with $\theta(0) = 0$ and $\bar{\theta}(0) = 1$, respectively. The functions A_{nk} are to be determined, and are easily seen to be given by

$$A_{nk} = \frac{2\,m/\hbar^2}{\psi_{n,-k}(z)\partial_z\psi_{nk}(z) - \psi_{nk}(z)\partial_z\psi_{n,-k}(z)}. \tag{6.45}$$

The particle-impurity scattering problem is now very easy to solve by means of the Lippmann–Schwinger equation for the scattering wave function $\Psi_{nk}(z)$, as

$$\Psi_{nk}(z) = \psi_{nk}(z) + \int_{-\infty}^{\infty} dz' G_0(E; z, z')H_I(z')\Psi_{nk}(z'). \tag{6.46}$$

For $H_I(z) = g\delta(z)$, we obtain

$$\Psi_{nk}(z) = \psi_{nk}(z) + gG_0(E; z, 0)\Psi_{nk}(0). \tag{6.47}$$

Setting $z = 0$ above, we calculate $\Psi_{nk}(0)$ as

$$\Psi_{nk}(0) = \frac{\psi_{nk}(0)}{1 - gG_0(E; 0, 0)}, \tag{6.48}$$

which, inserted back into equation (6.47) gives for the scattering state

$$\Psi_{nk}(z) = \psi_{nk}(z) + \frac{g\psi_{nk}(0)}{1 - gG_0(E; 0, 0)}G_0(E; z, 0). \tag{6.49}$$

In problem 1, the reader is asked to obtain the transmission and reflection amplitudes from the exact scattering state.

6.1.5 Impurity scattering in the tight-binding approximation

Let us consider now a single particle in a (hypercubic) tight-binding lattice, with first-quantized, position-represented Hamiltonian

$$H = -J\sum_{\langle ij\rangle}|i\rangle\langle j| + 2JD + U|0\rangle\langle 0| + \sum_{j\neq 0}V(j)|j\rangle\langle j|. \tag{6.50}$$

Above, $|i\rangle$ is the (multidimensional integer, $i \in \mathbb{Z}^D$) position of the particle within the lattice, $\langle ij\rangle$ restricts the first sum over nearest neighbours only and U and $V(j)$ represent the on-site and off-site potentials, i.e., the interactions with the impurity. A constant energy off-set of $+2JD$ has been added to the Hamiltonian so that the non-interacting ground state energy is zero.

The non-interacting part (that is, with $U = V(j) = 0$, $\forall j$) is easy to solve. The non-interacting states $|\psi\rangle$ are plane waves of the form

$$|\psi\rangle = \sum_{j\in\mathbb{Z}^D} e^{i\mathbf{k}\cdot j}|j\rangle, \tag{6.51}$$

and the eigenenergies are given by

$$\epsilon(\mathbf{k}) = -2J\sum_{i=1}^{D}[\cos(k_i) - 1], \tag{6.52}$$

where we have set the lattice spacing $d = 1$ for simplicity.

In our quest to simplify the problem, we set all off-site interactions to zero, that is, $V(j) = 0$ for all j while $U \neq 0$. The T-matrix is a constant in the quasi-momentum representation, and is easily obtained as

$$\frac{1}{t(z)} = \frac{1}{U} - \int_{\mathrm{BZ}} \frac{d^D k}{(2\pi)^D} \frac{1}{z - \epsilon(\mathbf{k})}. \tag{6.53}$$

In one and three spatial dimensions, some analytical solutions can be obtained. For $D = 1$, the integral in equation (6.53) for $z = \epsilon(k_0) + i\eta$ can be calculated using contour integration in the complex plane. To do this, we choose the following contour, oriented counterclockwise,

$$\Gamma = \Gamma_1 \bigcup \Gamma_2 \bigcup \Gamma_3 \bigcup \Gamma_4, \tag{6.54}$$

where $\Gamma_1 = (-\pi/d, \pi/d]$, $\Gamma_2 = \{\xi = \pi/d + is\colon s \in [0, R]\}$, $\Gamma_3 = \{\xi = s + iR\colon s \in (\pi/d, -\pi/d]\}$ and $\Gamma_4 = \{\xi = -\pi/d + is\colon s \in [R, 0]\}$. The limit $R \to \infty$ is taken at the end of the calculation, and it is easy to see that

$$\int_{\mathrm{BZ}} \frac{dk}{2\pi} \frac{1}{z - \epsilon(k)} = \oint_{\Gamma} \frac{d\xi}{2\pi} \frac{1}{z - \epsilon(\xi)}. \tag{6.55}$$

The integrand on the right-hand-side of equation (6.55) has two poles, located at $\xi_\pm = \pm k_0 \pm i\eta \csc(k_0)/2J$, with ξ_+ the only pole inside the contour. The residue, as $\eta \to 0^+$, is given by

$$\mathrm{Res}\left[\frac{1}{\epsilon(k_0) + i\eta - \epsilon(\xi)}\right]_{\xi=\xi_+} = -\frac{1}{2J \sin k_0}, \tag{6.56}$$

and, therefore

$$\int_{\mathrm{BZ}} \frac{dk}{2\pi} \frac{1}{z - \epsilon(k)} = -\frac{i}{2J \sin k_0}. \tag{6.57}$$

Hence, inserting equation (6.57) into equation (6.53), the T-matrix takes the form, for $z = \epsilon(k_0) + i\eta$,

$$\frac{1}{t(z)} = \frac{1}{U} + \frac{i}{2J \sin k_0}. \tag{6.58}$$

Before concluding the one-dimensional calculation in the momentum representation, we may also discuss bound states. Analytically continuing equation (6.58) to negative energies, by setting $k_0 = i\lambda$, $\lambda > 0$, and $E = \epsilon(i\lambda) = -2J(\cosh \lambda + 1)$ we obtain

$$\frac{1}{t(E)} = \frac{1}{U} + \frac{1}{2J \sinh \lambda}, \tag{6.59}$$

Therefore, the T-matrix has a pole for $U/J = -2\sinh\lambda$. Defining $\alpha = \exp(-\lambda)$, this is equivalent to

$$\frac{U}{J} = -\left[\frac{1}{\alpha} - \alpha\right]. \tag{6.60}$$

equation (6.60) above is analytically solvable, of course. It also presents a good surprise. However, this is easier to see in the position representation, and we shall not comment on it right now.

The one-dimensional problem in the position representation is very easy to solve, too. For bound states, we simply need to set

$$\psi(j) = \alpha^{|j|}, \tag{6.61}$$

and insert it into the stationary Schrödinger equation. We immediately see, by inspecting the Schrödinger equation at $j \neq 0$, that

$$E = -J\left(\frac{1}{\alpha} + \alpha - 2\right), \tag{6.62}$$

while for $j = 0$ we obtain

$$-2J\alpha + 2J + U = -J\left(\frac{1}{\alpha} + \alpha - 2\right). \tag{6.63}$$

This reduces to

$$\frac{U}{J} = -\left(\frac{1}{\alpha} - \alpha\right). \tag{6.64}$$

The above equation is, of course, identical to equation (6.60). However, in deriving equation (6.60) we implicitly assumed that $\alpha > 0$. For $U/J > 0$, there is actually a *normalizable*, that is, physically admissible solution with $\alpha < 0$ and, therefore, with energy *above* the band (see equation (6.62)), which implies that there can be repulsively bound states with $E > 2J$. This is due to the fact that there are no available continuum states a potential bound state may decay into, and therefore can be stable. Note that, in a more realistic lattice model, these correspond to in-gap bound states, which can be seen to exist as well. As for the scattering states, these take the form

$$\psi(j) = \sin(k|j| + \theta_k), \tag{6.65}$$

which are clearly solutions to the stationary Schrödinger equation for $j \neq 0$. Inserting equation (6.65) into the Schrödinger equation at $j = 0$, we obtain the following for the phase shifts θ_k

$$\cot\theta_k = \frac{U}{2J\sin k}. \tag{6.66}$$

We now study the three-dimensional case, which has a slightly more interesting phenomenology. The position representation is not particularly useful for analytical calculations, but the momentum representation does offer some analytical insight. In particular, for $D = 3$ there is a zero energy resonance, which can be calculated analytically thanks to Watson's integrals [1]. The one we need, in particular, is the following

$$\begin{aligned} W &= \frac{1}{\pi^3}\int_0^\pi dx \int_0^\pi dy \int_0^\pi dz \frac{1}{3 - \cos x - \cos y - \cos z} \\ &= \frac{\sqrt{6}}{96\pi^3}\Gamma\left(\frac{1}{24}\right)\Gamma\left(\frac{5}{24}\right)\Gamma\left(\frac{7}{24}\right)\Gamma\left(\frac{11}{24}\right) = 0.505\,462\ldots. \end{aligned} \tag{6.67}$$

The integral in equation (6.53), for $D = 3$ and at zero energy ($z = 0$), is obviously related to W in equation (6.67), as

$$\int \frac{d^3k}{(2\pi)^3}\frac{1}{-\epsilon(\mathbf{k})} = -\frac{W}{2J}, \tag{6.68}$$

which implies that the resonance position is located at $U = U_c$, with

$$U_c/J = -3.9567\ldots. \tag{6.69}$$

Therefore, there is a bound state in three spatial dimensions if $U < U_c$, and none otherwise.

The position of the resonance has further implications regarding the continuum limit. As is usual in lattice theories, the continuum limit only exists around criticality [2]. For this, we should investigate the scattering length, and restore the lattice spacing d (which should tend to zero to achieve a continuum limit). Before doing that, let us take a look at the on-shell T-matrix at zero energy ($z = 0$). From equation (6.53), and equation (6.68), we have

$$\frac{1}{t(0)} = \frac{1}{U} + \frac{W}{2J}. \tag{6.70}$$

Clearly, from dimensional analysis we must have the following dependence on the lattice spacing d,

$$\frac{1}{t(0;\, d)} = \frac{1}{U(d)d^3} + \frac{W}{2J(d)d^3}, \tag{6.71}$$

where we have made the d-dependence of every quantity explicit. The dependence of the tunnelling rate J on the spacing is very simple, since it comes directly from the finite-difference expression for the second derivative, see equation (4.136) and the subsequent discussion. It takes the explicit form

$$J(d) = \frac{\hbar^2}{2md^2}. \tag{6.72}$$

As the lattice spacing is taken to zero, $d \to 0^+$, equation (6.71) should take the limiting value

$$\lim_{d\to 0^+} t(0; d) = \frac{2\pi\hbar^2}{m}a, \tag{6.73}$$

where a is the scattering length. Therefore, the only possibility is to have, as $d \to 0^+$,

$$U = U_c\left[1 - \frac{1}{2\pi W}\frac{d}{a} + O(d^2)\right], \quad d \to 0^+. \tag{6.74}$$

That is, U/J is finite, and U must be linearly close to its critical value as the lattice spacing is taken to zero. In fact, this point is crucial in order to set the scattering length to a finite value. We will see that this is not exclusive to three dimensions, but is also required in one dimension with a different interaction potential.

We can also show that we are doing everything correctly regarding the continuum limit by taking $U = U_c$ and diagonalizing the Hamiltonian with several values of L, which are fixed, and varying lattice spacing d (therefore having a varying number of lattice sites $N_s = (L/d)^3$). In figure 6.1, we plot $EmL^2/\hbar^2$, with E the ground state energy, as a function of d/L for several values of d, as a function of $d/L = N_s^{-1/3}$. Being on resonance, all curves collapse into a single curve. The expected result, from the continuum limit, is given by equation (2.119), with m replaced with $2m$.

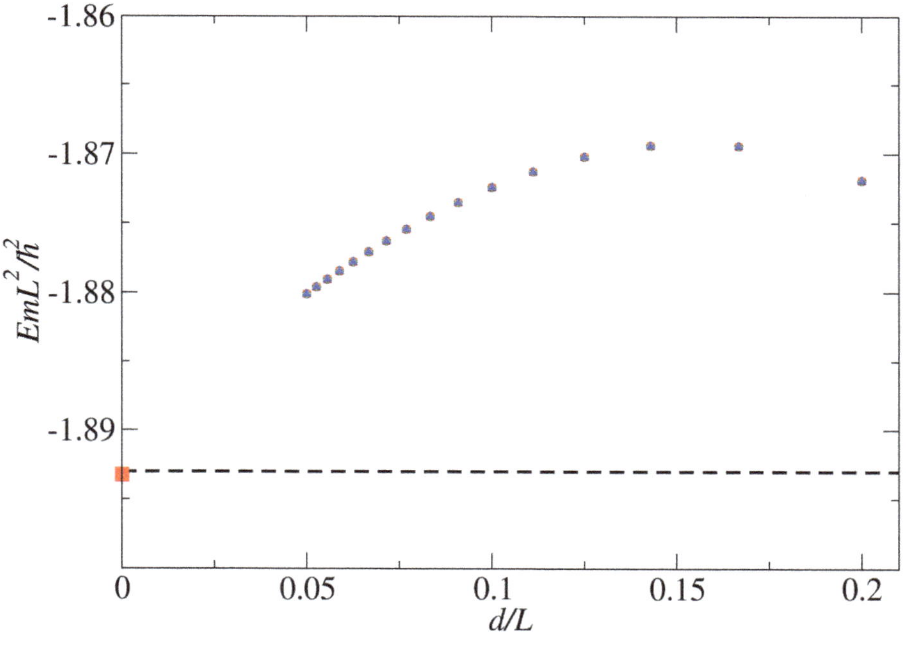

Figure 6.1. Dots: dimensionless ground state energy times L^2 as a function of d/L for several numerical implementations. The red square at $d/L = 0$ is the extrapolated result, while the dashed line is the result extracted from equation (2.119) with m replaced with $2m$.

Extrapolating the results, using a second degree polynomial in d, to $d = 0^+$ (the continuum limit), we obtain a very accurate result.

Let us now take a look at criticality and the continuum limit for the one-dimensional model in the odd-wave channel (antisymmetric states), with $V(1)$ finite and $V(|j| > 1) = 0$. The scattering states take the form

$$\psi(j) = \text{sgn}(j)\sin(k|j| + \theta_k), \tag{6.75}$$

with $\text{sgn}(j)$ the signum function with $\text{sgn}(0) \equiv 0$. (Note that, unlike in the continuum where signum is a distribution, here it is a usual function, and we must assign it a value at $j = 0$. This must be zero since we are working with the odd-wave channel.) Introducing equation (6.77) into the stationary Schrödinger equation, we obtain for the phase shifts

$$-\sin(kd)\cot\theta_k = \frac{J}{V} + \cos(kd). \tag{6.76}$$

Near the continuum limit, and as $k \to 0$, the above expression becomes

$$-k\cot\theta_k \to \frac{1}{a} = \frac{J}{Vd} + \frac{1}{d} + O(d^2). \tag{6.77}$$

Therefore, we must have

$$\frac{V}{J} = -\frac{1}{1 - d/a}. \tag{6.78}$$

Since the scattering length becomes infinite on resonance, then V/J must be tuned to near-resonant as $d \to 0^+$, that is, $V/J \approx -1$, if we are to take the continuum limit.

6.1.6 Edge modes in graphene

The description of graphene that we have already used both in the tight-binding lattice representation as well as in its effective low-energy description in the continuum has so far only been studied in the infinite size limit and without boundaries. Square lattices do not show any important surface effects, besides convenient quantization conditions that allow for scattering properties to be obtained, as we have seen already. However, the structure of graphene is non-trivial, and surface effects can be important. Here, we study such effects within its continuum low-energy description.

A finite or semi-infinite honeycomb lattice has two types of regular edges, depending on how the system is terminated. These are called zig-zag and armchair edges, respectively, and are pictured in figure 6.2. Armchair edges are not particularly special, but zig-zag edges do support (topologically trivial) edge states. The difference between these two types of edges lies in the boundary conditions. In particular, dynamical states of the system with a zig-zag edge on the left, and no ends on the right (i.e., a semi-infinite system), need to be supplemented with the boundary condition,

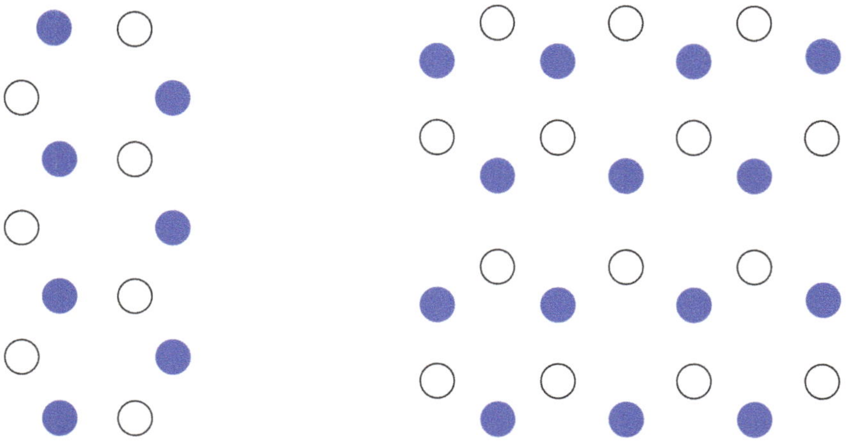

Figure 6.2. Left: Zig-zag left edge termination in a honeycomb lattice. Right: Armchair left edge termination in the same lattice. The A and B sublattices are represented with open and filled dots, respectively.

$$\phi_b(0, y) = 0, \tag{6.79}$$

while for an armchair end on the left, the boundary condition is

$$\phi_a(0, y) = \phi_b(0, y) = 0. \tag{6.80}$$

Note that, unlike for zig-zag edges, for which we may disregard coupling between the two Dirac points $\mathbf{K}$ and $\mathbf{K}'$, in the case of armchair edges, these necessarily couple because otherwise we have one too many boundary conditions. Let us solve for the edge states with a zig-zag edge on the left. Since we require the edge state to be bound to the edge, it should have the form

$$\psi(x, y) = e^{-\lambda x}\phi(x, y), \tag{6.81}$$

where $\lambda > 0$ and ϕ plane-wave-like. Note that, since the system has only an edge on the left, the wave function in equation (6.81) is normalizable in the x-direction. Now, since the system is only translationally invariant in the y-direction, the spinor ψ^{T} can be written as

$$(\psi_a(x, y), \psi_b(x, y)) = e^{ik_y y}e^{-\lambda x}(\phi_a, \phi_b), \tag{6.82}$$

and the Schrödinger equation takes the form

$$-iv_F(\partial_x + k_y)\phi_b e^{-\lambda x} = E\phi_a e^{-\lambda x}, \tag{6.83}$$

$$-iv_F(\partial_x - k_y)\phi_a e^{-\lambda x} = E\phi_b e^{-\lambda x}. \tag{6.84}$$

The boundary condition on ϕ_b, equation (6.79), implies that $\phi_b = 0$. Therefore, equation (6.83) implies that $E = 0$, and from equation (6.84), we obtain

$$\lambda = -k_y, \tag{6.85}$$

which yields a normalizable state if $k_y < 0$. This is an artifact of having considered only the low-energy physics around $\mathbf{K}$. If we consider this around $\mathbf{K}'$, then the state is normalizable for $k_y > 0$, and there is a surface state for all k_y.

6.1.7 Topological edge modes in the Hofstadter model

There are systems that have some interesting features, including chiral edge modes (which move along a single direction), called 'topologically non-trivial'. Here, we will study a particular case of these edge states, without even mentioning any topological properties (besides the edge modes themselves). These, as important as they are, can be consulted elsewhere [3], and here we focus on the one-body bound state-like states that these edge states represent in a particular model.

The model we study is the so-called Hofstadter model on a two-dimensional square tight-binding lattice. The model represents a discretized version of the Hamiltonian for one electron (or any other charged particle) under the influence of a constant magnetic field. Some preliminary concepts will be useful here. Recall that a constant magnetic field $\mathbf{B} = B_0\hat{z}$ perpendicular to the $x - y$ plane is obtained from a vector potential $\mathbf{A}(\mathbf{r})$ such that

$$B_0\hat{z} = \nabla \times \mathbf{A}(\mathbf{r}). \tag{6.86}$$

There are infinitely many different functions $\mathbf{A}(\mathbf{r})$ that produce the field in equation (6.86), corresponding to different choices of gauge. The one we are interested in, because of its simplicity on a lattice, is the so-called Landau gauge, wherein

$$\mathbf{A}(\mathbf{r}) = B_0 x\hat{x}. \tag{6.87}$$

The Hamiltonian for a non-relativistic electron in this gauge is given by

$$H = \frac{1}{2m}\left(\mathbf{p} - \frac{e}{c}\mathbf{A}(\mathbf{r})\right)^2. \tag{6.88}$$

If we are interested in a two-dimensional system, then the magnetic field is actually a scalar, since the two-dimensional cross-product is scalar. The Hamiltonian is given in this case by equation (6.88) with $\mathbf{p}$ the two-dimensional momentum operator (p_x, p_y).

To place the system on a lattice, we introduce the so-called Peierls substitution. Without a magnetic field, it consists of elevating the tight-binding energy dispersion $\epsilon(\mathbf{k})$ to an operator, in the following way

$$\hat{\epsilon} \equiv \epsilon(\hat{\mathbf{k}}), \tag{6.89}$$

where

$$\hat{\mathbf{k}} = -i\nabla_{\mathbf{r}}. \tag{6.90}$$

Since the lattice is two-dimensional and square, the simplest tight-binding dispersion (from nearest-neighbour tunnelling only) is

$$\epsilon(\mathbf{k}) = -2J\left[\cos(k_x d) + \cos(k_y d)\right], \tag{6.91}$$

where we have omitted any constant off-set that does not affect the dynamics in any way. Using the Peierls substitution, we have

$$\cos(\hat{k}_x d) = \frac{1}{2}[e^{i\hat{k}_x d} + e^{-i\hat{k}_x d}] = \frac{1}{2}[e^{d\partial_x} + e^{-d\partial_x}]. \tag{6.92}$$

The action of the above operators on a function f defined on a one-dimensional lattice $\{x = nd\colon n \in \mathbb{Z}\}$ is obtained as follows

$$\exp(\pm d\partial_x)f(x) = \sum_{j=1} \frac{(\pm 1)^j}{j!} d^j \frac{\partial^j f}{\partial x^j}(x) = f(x \pm d). \tag{6.93}$$

The last equality above can be readily checked by expanding $f(x \pm d)$ around $d = 0$. Equation (6.93) is quite familiar, since the operators are just translation operators by a constant d. Hence, the action of $\epsilon(\hat{\mathbf{k}})$ on a wave function defined on a square lattice is just nearest-neighbour hopping with tunnelling rate J. A non-trivial Peierls substitution is used when the system is under the influence of a magnetic field. In that case, one replaces

$$\mathbf{p} \leftrightarrow \hbar\hat{\mathbf{k}} - (e/c)\mathbf{A}. \tag{6.94}$$

The dispersion, as an operator, becomes the Hamiltonian

$$H = -2J\left[\cos(\hat{k}_x d - a_x(\mathbf{r})) + \cos(\hat{k}_y d - a_y(\mathbf{r}))\right], \tag{6.95}$$

where we have defined $\mathbf{a}(\mathbf{r})$ by regrouping constants in order to avoid overloading the notation. The Hamiltonian in equation (6.95) can have a quite non-trivial position representation. However, for the problem at hand and in the Landau gauge, it is quite simple. Since $\mathbf{A}_x = 0$ and $\mathbf{A}_y = \mathbf{A}_y(x)$, the commutator $[\hat{k}_y, a_y(x)]$ vanishes, and in the second quantization, the Hamiltonian (6.95) takes the form (setting $d = 1$)

$$H(k_y) = -J\sum_{j_x}\left[c^\dagger_{j_x+1}c_{j_x} + \text{H. c.}\right] - 2J\sum_{j_x}\cos\left(2\pi j_x/\tau - k_y\right)c^\dagger_{j_x}c_{j_x}. \tag{6.96}$$

Above, τ is the period of the tunnelling in the y-direction, where we have set $a_y(x) = 2\pi x/(\tau d)$ and, with $d = 1$, $x = j_x$ and $\tau d = \tau$.

We now diagonalize Hamiltonian $H(k_y)$ in equation (6.97) with open boundary conditions and size (in the x-direction) $L_s = \tau - 1$, for $\tau = 10$. Every state in the system is then an edge mode, since edge modes in the infinite size limit are of the generalized Bloch form

$$\psi(j_x) = \phi_\alpha(j_x)\alpha^x, \tag{6.97}$$

with $|\alpha| < 1$ and $\phi_\alpha(j_x + \tau) = \phi_\alpha(j_x)$. Since the exponential part of the generalized Bloch wave function is non-zero, then the boundary condition at $j_x = 0, \psi(0) = 0$, is satisfied only if $\phi_\alpha(0) = 0$. Since ϕ_α is periodic with period τ, then the boundary

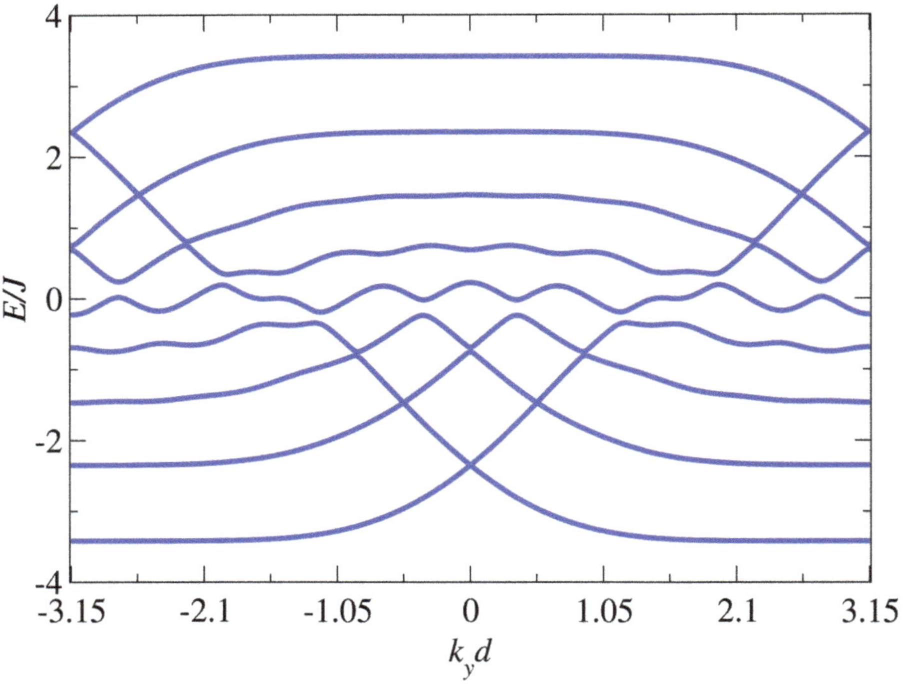

Figure 6.3. Energy of the edge states for the Hofstadter model with $\tau = 10$ as a function of k_y.

condition at the origin implies that $\phi_\alpha(\tau) = 0$. Hence, we get all the edge mode energies by diagonalizing the system in a box of size $\tau - 1$. In figure 6.3, we plot the edge state energies for $\tau = 10$ obtained in this way.

6.1.8 Topological edge modes in a Chern insulator

Another interesting system where edge modes are present is a so-called Chern insulator. These are systems that have non-trivial topology, and chiral edge states, but no net magnetic field. Here, we study and calculate the edge modes of the simplest of these models on a square lattice.

The particular model we are interested in has two states per lattice site. Its Hamiltonian, in the quasi-momentum representation, is given by the 2×2 Hamiltonian in two spatial dimensions [4]

$$H = \epsilon(\mathbf{k}) + \mathbf{d}(\mathbf{k}) \cdot \vec{\sigma}, \tag{6.98}$$

where $\vec{\sigma} = (\sigma_x, \sigma_y, \sigma_z)$ is the vector of spin-1/2 Pauli matrices, while ϵ and $\mathbf{d}$ are given, respectively, by (we set the lattice spacing $d = 1$)

$$\epsilon(\mathbf{k}) = C - 2D\big(2 - \cos k_x - \cos k_y\big), \tag{6.99}$$

$$\mathbf{d}(\mathbf{k}) = (A \sin k_x, A \sin k_y, M(\mathbf{k})), \tag{6.100}$$

with

$$M(\mathbf{k}) = M + B(\cos k_x + \cos k_y). \tag{6.101}$$

The normalized eigenstates of Hamiltonian (6.98) are denoted $\psi_s(\mathbf{k}) = (a_s(\mathbf{k}), b_s(\mathbf{k}))^{\mathrm{T}} \exp(i\mathbf{k} \cdot \mathbf{r})$, which the reader should be able to readily calculate (see problem 2).

To obtain the edge modes, we could transform the Hamiltonian to the position representation, and use exact diagonalization on the lattice. This would be done using the Peierls substitution, which was thoroughly explained in the previous subsection. However, it is simpler to use quasi-momentum space methods, which allow for the calculation of the edge modes in semi-infinite space very simply. To do this, we place a hard wall (open boundary condition) at $x = 0$. This is equivalent to adding a line potential of infinite strength located at $x = 0$, that is, a local potential of the form

$$\langle x'y'|V|xy\rangle = \Gamma\delta_{x,0}\delta_{x,x'}\delta_{y,y'}, \quad \Gamma \to \infty. \tag{6.102}$$

The advantage of such an approach is that the Fourier transform of the above potential is trivial. Note that, by imposing the boundary condition $\psi(0, y) = 0$, we will have edge modes attached to the right *and* left of the boundary, with opposite chirality.

Let us set up now the Lippmann–Schwinger equation for the residue of the T-matrix at the edge mode energies for fixed, conserved k_y. If this is called τ, it satisfies

$$\tau = VG_0(z, k_y)\tau, \tag{6.103}$$

where $\tau \neq \tau(z)$ (recall that the residue is energy-independent), and the non-interacting Green's function is given by

$$G_0(z, k_y) = \left[z - H(\hat{k}_x, k_y)\right]^{-1}. \tag{6.104}$$

Since an edge can be viewed as a line interaction, for fixed k_y (which we implicitly assume to be the case for simplicity of notation), we have, in the momentum representation,

$$\begin{aligned}\langle s'k_x'|V|sk_x\rangle &= \Gamma(a_{s'}^*(k_x', k_y), b_{s'}^*(k_x', k_y))(a_s(k_x, k_y), b_s(k_x, k_y))^{\mathrm{T}} \\ &= \Gamma\left[a_{s'}^*(k_x', k_y)a_s(k_x, k_y) + b_{s'}^*(k_x', k_y)b_s(k_x, k_y)\right], \quad \Gamma \to \infty.\end{aligned} \tag{6.105}$$

Inserting equation (6.105) into the Lippmann–Schwinger equation, equation (6.103), in the momentum representation, we obtain

$$\begin{aligned}\langle s'k_x'|\tau|sk_x\rangle = \Gamma\sum_{\sigma=\pm}\int_{-\pi}^{\pi}\frac{dq_x}{2\pi}\frac{a_{s'}^*(k_x', k_y)a_\sigma(q_x, k_y) + b_{s'}^*(k_x', k_y)b_\sigma(q_x, k_y)}{E - E_\sigma(q_x, k_y)}\times \\ \times\langle\sigma q_x|\tau|sk_x\rangle.\end{aligned} \tag{6.106}$$

It is straightforward to see that the matrix elements of the residue take the form

$$\langle s'k_x|\tau|sk_x\rangle = \tau^a a_{s'}^*(k_x', k_y) + \tau^b b_{s'}(k_x', k_y), \tag{6.107}$$

where τ^a and τ^b are constant coefficients, independent of s, s' and the momentum k_x'. Inserting equation (6.107) into equation (6.106), we obtain the following system of coupled equations

$$\tau^a = \Gamma \sum_{\sigma=\pm} \int_{-\pi}^{\pi} \frac{dq_x}{2\pi} \frac{a_\sigma(q_x, k_y)}{E - E_\sigma(q_x, k_y)} \left[a_\sigma^*(q_x, k_y)\tau^a + b_\sigma^*(q_x, k_y)\tau^b \right], \tag{6.108}$$

$$\tau^b = \Gamma \sum_{\sigma=\pm} \int_{-\pi}^{\pi} \frac{dq_x}{2\pi} \frac{b_\sigma(q_x, k_y)}{E - E_\sigma(q_x, k_y)} \left[a_\sigma^*(q_x, k_y)\tau^a + b_\sigma^*(q_x, k_y)\tau^b \right]. \tag{6.109}$$

We now define, given two functions f_σ and g_σ,

$$I_{fg}(E, k_y) = \sum_{\sigma=\pm} \int_{-\pi}^{\pi} \frac{dq_x}{2\pi} \frac{f_\sigma^*(q_x, k_y) g_\sigma(q_x, k_y)}{E - E_\sigma(q_x, k_y)}. \tag{6.110}$$

With this, solving equations (6.108) and (6.109) results in the following equation for the bound state energy (with fixed k_y)

$$1 = \frac{|I_{ab}(E, k_y)|^2}{\left[1/\Gamma - I_{bb}(E, k_y)\right]\left[1/\Gamma - I_{aa}(E, k_y)\right]} \to 1 = \frac{|I_{ab}(E, k_y)|^2}{I_{bb}(E, k_y) I_{aa}(E, k_y)}, \quad \Gamma \to \infty. \tag{6.111}$$

Note that, since we have only prevented (by means of taking $\Gamma \to \infty$) $x = 0$ from being populated, equation (6.111) will have two in-gap solutions for every k_y (of equal energy), one of them non-zero for $x > 0$ and the other one non-zero for $x < 0$, that have opposite chirality. In figure 6.4, we plot one of the edge bands as obtained from equation (6.111) for a case with non-trivial topology ($M/A = 3/4$).

6.1.9 An impurity in a Chern insulator

We continue studying the same system as in the previous subsection. Here, instead of focusing on edge states, we study the effect of a static impurity located at an arbitrary point of the lattice. We consider the potential generated by the impurity to be zero range and of strength U. The matrix elements of the interaction in the momentum representation take the form

$$\langle s'\mathbf{k}'|V|s\mathbf{k}\rangle = U[a_{s'}^*(\mathbf{k}')a_s(\mathbf{k}) + b_{s'}^*(\mathbf{k}')b_s(\mathbf{k})]. \tag{6.112}$$

The derivation of the bound state energy equation is done analogously to that for the edge states, with the difference that now the problem is two-dimensional and neither k_x nor k_y are conserved. It reads

$$\frac{1}{U} = \sum_{\sigma=\pm} \int_{\mathrm{BZ}} \frac{d\mathbf{q}}{(2\pi)^2} \frac{\xi(U, E) a_\sigma^*(\mathbf{q}) b_\sigma(\mathbf{q}) + |b_\sigma(\mathbf{q})|^2}{E - E_\sigma(\mathbf{q})}, \tag{6.113}$$

where ξ is given by

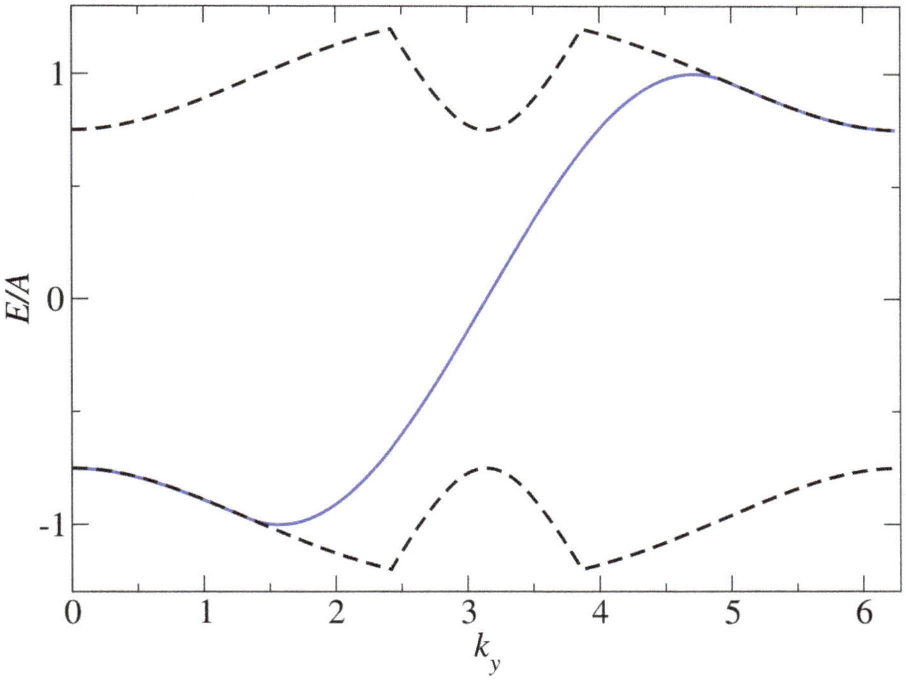

Figure 6.4. Energy of chiral edge states with $M/A = 3/4$ and $B/A = 1$ as a function of k_y (blue solid line). For comparison, the highest ($s = -$) and lowest ($s = +$) bulk state energies are also plotted (black dashed line).

$$\xi(U, E) = \frac{\sum_{\sigma=\pm} \int_{\mathrm{BZ}} \frac{d\mathbf{q}}{(2\pi)^2} \frac{a_\sigma(\mathbf{q}) b_\sigma^*(\mathbf{q})}{E - E_\sigma(\mathbf{q})}}{1/U - \sum_{\sigma=\pm} \int_{\mathrm{BZ}} \frac{d\mathbf{q}}{(2\pi)^2} \frac{|a_\sigma(\mathbf{q})|^2}{E - E_{\sigma(\mathbf{q})}}}. \tag{6.114}$$

The most interesting feature exhibited by the in-gap bound state is the existence of bound states even for a vacancy, that is, for $U \to \infty$.

6.1.10 Off-resonant edge-impurity collisions in a Chern insulator

We now combine the settings of the last two sections in the following way. We consider a Chern insulator with one edge (at $x = 0$, for instance), and a zero range impurity (which can also be a vacancy) located near the edge of the sample. Clearly, since without the impurity the energies of the edge states cross the entire band gap, and without an edge the bound state lies within the band gap, once an edge and an impurity are in place at the same time, there can be no actual bound state. Indeed, the bound state hybridizes with the edge states, even strongly for energies near that of the bound state, to produce a resonance.

It is instructive to study the naïve elastic scattering problem of a chiral state with a static impurity within an EFT of sorts. Consider an up-moving chiral state with speed v_y, and low-energy non-interacting Hamiltonian

$$H_0 = -i\hbar v_y \partial_y. \tag{6.115}$$

Insisting on modelling a near-edge-impurity located around $y = 0$ (and small $|x|$) as a potential, we may choose the simplest of all, namely a Dirac delta function, so that the full elastic Hamiltonian takes the form

$$H = -i\hbar v_y \partial_y + g_{ee}\delta(y), \tag{6.116}$$

where g_{ee} is the strength of the interaction in the effective edge (e) channel. In the infinite size limit, the scattering states (bound states cannot exist, obviously) are extremely easily calculated. Writing down the stationary Schrödinger equation with energy $E = \hbar v_y k_y$, we have

$$-i\hbar v_y \psi'(y) + g_{ee}\delta(y)\psi(y) = \hbar v_y k_y \psi(y), \tag{6.117}$$

which means that the logarithmic derivative of the scattering state must yield

$$\frac{\psi'(y)}{\psi(y)} = i\left[k_y - \frac{g_{ee}}{\hbar v}\delta(y)\right]. \tag{6.118}$$

The solution for a particle incident from below ($y < 0$) as a plane wave is given by

$$\psi(y) = e^{ik_y y}\exp\left(-i\frac{g_{ee}}{\hbar v}\theta(x)\right), \tag{6.119}$$

where $\theta(x)$ is the Heaviside step function. Therefore, the scattering states simply suffer a sudden phase change at $x = 0$. Actually, regardless of the potential $V(y)$, which may be arbitrary, all the impurity does is add a phase difference between the incident ($y \to -\infty$) and outgoing ($y \to \infty$) wave. This is due to the fact that an isolated chiral state has no time-reversed partner into which scatter and, therefore, it must continue unaltered when elastically scattering with an impurity.

The question we need to answer now is whether, at least for far off-resonant edge states (with energies far from the bound state energy), the scattering states of the chiral edge states off the impurity, with the bulk fully taken into account, are well modelled by equation (6.119). The answer is negative. The issue now is that studying the full problem within scattering theory is quite complicated. We can work around this by using finite-size diagonalization, with periodic boundary conditions in y (size L_y) and open boundary conditions in x (size L_x). If the model in equation (6.119) is accurate, then we expect the quantization of k_y to be

$$\psi(0) = \psi(L_y) \Rightarrow 1 = e^{ik_y L_y}e^{-ig_{ee}/\hbar v_y} \Rightarrow k_y = \frac{2\pi n}{L_y} + \frac{g_{ee}}{\hbar v_y L_y}, \tag{6.120}$$

where n is an integer. If at all, the quantization should be valid for energies near $E = 0$, where the dispersion is most accurately linear. Equation (6.120) predicts an equal shift of the edge state energies, all in the same direction, given by

$$\Delta E_{\text{el}} = \frac{g_{ee}}{L_y}. \tag{6.121}$$

Exact diagonalization of the lattice Hamiltonian (see problem 3) reveals an inadequacy of the above theory, and in particular of equation (6.121), to describe the spectrum. It is necessary, but beyond the scope of this section, to consider a model that takes into account other terms as well as the resonance to properly describe edge state-impurity collisions in this setting.

6.1.11 Single photon in an array of cavities

In section 4.3, we derived the Hamiltonian of a photon–atom system in a cavity, equation (4.153), under a number of assumptions. In particular, we used the rotating-wave approximation, which allows us to simplify the interacting term H_I as

$$H_I = \hbar g(\sigma_- + \sigma_+)(a + a^\dagger) \approx \hbar g(\sigma_+ a + a^\dagger \sigma_-), \tag{6.122}$$

whenever the system is in the weak-coupling regime, that is, for small g/ω_{eg}. The clear theoretical and computational advantage of the RWA interaction in the approximation in equation (6.122) is given by the fact that the Jaynes–Cummings Hamiltonian (4.153) conserves the total number of excitations. That is, if N is the number of photons in the cavity and $s_z = 0$ if the atom is in the ground state and $s_z = 1$ otherwise, then $N + s_z$ is conserved. This allows for generalizations in which some interesting physics can be calculated exactly. We do this in the following.

Consider an infinite, equally spaced, one-dimensional array of coupled cavities, such that a photon can tunnel from a cavity to its neighbours with rate $-J$. Assume that there is one and only one two-level atom located in a particular cavity, which we choose to be the central one, labelled $j = 0$, without loss of generality. The coupled Hamiltonian consists of a nearest-neighbour tunnelling term for the photons, an energy off-set for the photons ($\hbar\omega$) together with a Jaynes–Cummings interaction at site $j = 0$, that is

$$H = \sum_j \left[\hbar\omega a_j^\dagger a_j + J(a_{j+1}^\dagger a_j + \text{H. c.})\right] + \frac{1}{2}\hbar\omega_{eg}\sigma_z + \hbar g(\sigma_+ a_0 + \sigma_- a_0^\dagger). \tag{6.123}$$

We begin with the ground state, $|G\rangle$, of the system, which is extremely simple, and is given by the vacuum of photons ($|0\rangle$) and the atom in the ground state, that is

$$|G\rangle = |\downarrow\rangle|0\rangle. \tag{6.124}$$

The reader can easily verify that $|G\rangle$ is an eigenstate of H in equation (6.124), and that it has the lowest energy.

The first non-trivial problem we can think of consists of studying the subspace with $N + s_z = 1$. In particular, we study the scattering and bound states of one photon off the atomic, number-non-conserving impurity. The states $|\psi\rangle$ take the form

$$|\psi\rangle = \sum_j \phi(j) a_j^\dagger |0\rangle|\downarrow\rangle + \alpha|\uparrow\rangle|0\rangle, \tag{6.125}$$

where $\phi(j)$ will be called the photon wave function, while α is a c-number representing the amplitude of the atomic excitation. We solve the stationary Schrödinger equation for $|\psi\rangle$ in equation (6.125) with Hamiltonian in equation (6.123), $H|\psi\rangle = E|\psi\rangle$. Projecting the resulting equation onto the $\langle \downarrow |$ and $\langle \uparrow |$ subspaces, we obtain the following two coupled equations

$$\left(\hbar\omega - \frac{\Delta}{2}\right)\phi(j) + J(\phi(j+1) + \phi(j-1)) + \hbar g\alpha\delta_{j0} = E\phi(j), \tag{6.126}$$

$$\hbar g\phi(0) + \alpha\frac{\Delta}{2} = E\alpha, \tag{6.127}$$

where we have defined $\Delta = \hbar\omega_{eg}$. To solve equations (6.126) and (6.127), we eliminate the amplitude α in equation (6.127), and introduce it into equation (6.126), obtaining the following equation for the photonic wave function $\phi(j)$

$$\left(\hbar\omega - \frac{\Delta}{2}\right)\phi(j) + J(\phi(j+1) + \phi(j-1)) + \frac{(\hbar g)^2}{E - \frac{\Delta}{2}}\delta_{j0}\phi(0) = E\phi(j). \tag{6.128}$$

Note how the above Schrödinger equation takes exactly the same form as that for one particle scattering off an impurity with (energy-dependent) zero range potential, that was solved with a different method in section 6.1.5. The reader is encouraged to solve equation (6.128) for both scattering and bound states.

6.2 Two-body problems

In this section we study a number of two-body problems, most of which feature phenomenology not present in the typical relative coordinate separated, one-body version. In particular, we study two-magnon collisions in a magnet, the two-boson problem in a harmonic trap, and the two-body problem on a genuine periodic potential in one spatial dimension.

6.2.1 Two-magnon scattering

Here, we consider the scattering of two elementary magnetic excitations—known as magnons—in a ferromagnet. In principle, these excitations exist on top of a many-body background (the ferromagnetic ground state) but, at least for the model we shall study, the n-magnon problem with an infinite number of spins ($N \to \infty$) behaves exactly as a standard n-body problem on a lattice.

The particular model we are interested in is the isotropic, spin-1/2 Heisenberg model, with Hamiltonian

$$H = J\sum_{\langle i,j\rangle}\mathbf{S}_i \cdot \mathbf{S}_j, \tag{6.129}$$

where $\mathbf{S}_i = (S_i^x, S_i^y, S_i^z)$ is the spin-1/2 operator. To make it more convenient, we re-express equation (6.129) in terms of the Pauli matrices, using $\mathbf{S}_i = (\hbar/2)\vec{\sigma}_i$. Also, choosing the quantization axis as z, and defining the ladder Pauli matrices $\sigma^{\pm}$, via

$$\sigma_i^{\pm} = \sigma_i^x \pm i\sigma_i^y, \tag{6.130}$$

we rewrite Hamiltonian (6.129) as

$$H = \frac{J\hbar^2}{8} \sum_{\langle i,j \rangle} \left[\sigma_i^+ \sigma_j^- + \sigma_i^- \sigma_i^+ + 2\sigma_i^z \sigma_j^z \right]. \tag{6.131}$$

If $J < 0$, we can prove that the ground state $|G\rangle$ of H is fully ferromagnetic, that is,

$$|G\rangle = \otimes_{i=1}^N | \uparrow \rangle_i, \tag{6.132}$$

where $| \uparrow \rangle_i$ is the eigenstate of σ_i^z with eigenvalue +1. Note that the ground state is not unique, since the state $|G'\rangle$ with all spins down is also a ground state of the system. The actual physical system will be in one or the other ground state, and not a combination of both, by means of spontaneous symmetry breaking. We will consider, without loss of generality, that the ground state is given by equation (6.132). Inserting equation (6.132) into the stationary Schrödinger equation, it is immediate to see that it is an eigenstate with eigenenergy

$$E_0 = \frac{J\hbar^2}{4} Nz, \tag{6.133}$$

where z is the coordination number of the lattice (see section 4.2.1).

We obtain first the single-spin excitations over the ground state, particularizing to one spatial dimension. This is possible since $S^z = \sum_i S_i^z$ is conserved, that is, $[H, S^z] = 0$. The basis of single-spin excitations are given by states of the form

$$|j\rangle \equiv \sigma_i^- |G\rangle, \tag{6.134}$$

and we shall seek one-magnon eigenstates as linear combinations of the form

$$|\psi\rangle = \sum_{\ell} \psi_\ell |\ell\rangle. \tag{6.135}$$

Introducing equation (6.135) into the stationary Schrödinger equation, we obtain the following difference equation for the amplitudes of the wave function,

$$\frac{J\hbar^2}{2} \left[\psi_{j+1} + \psi_{j-1} \right] = \Delta E \psi_j, \tag{6.136}$$

where $\Delta E = E - E_0$ is the excitation energy with respect to the ground state. Clearly, equation (6.136) is solved by a plane wave, namely

$$\psi_j = e^{ikj}, \tag{6.137}$$

with excitation energy

$$\Delta E(k) \equiv \epsilon(k) = J\hbar^2 \cos(k). \tag{6.138}$$

The two-magnon subspace is spanned instead by the following basis

$$|j\ell\rangle = \sigma_j^- \sigma_\ell^- |G\rangle, \quad j < \ell. \tag{6.139}$$

Note that, due to the commutation relations (hard-core bosonic) between the ladder Pauli matrices, $j < \ell$ is chosen to avoid double counting of states. The eigenstates take the form

$$|\psi\rangle = \sum_{j<\ell} \psi_{j\ell} |j\ell\rangle. \tag{6.140}$$

We define the 'first-quantized' wave function $\phi_{j\ell}$, where j and ℓ are not necessarily ordered (that is, they can take any value $j \neq \ell$), such that the actual amplitudes $\psi_{j\ell}$ are given by

$$\psi_{j\ell} = \frac{1}{\sqrt{2}}\left[\phi_{j\ell} + \phi_{\ell j}\right]. \tag{6.141}$$

It is straightforward to see that $\phi_{j\ell}$ satisfies the following stationary Schrödinger equation

$$\frac{J\hbar^2}{2}\left[\phi_{j+1,\ell} + \phi_{j-1,\ell} + \phi_{j,\ell+1} + \phi_{j,\ell-1}\right] + J\hbar^2 \delta_{|j-\ell|,1} \phi_{j,\ell} = \Delta E \phi_{j,\ell}, \tag{6.142}$$

with the further conditions that $\phi_{j\ell} = \phi_{\ell j}$ (symmetric) and $\phi_{jj} = 0$ (hard-core). Clearly, the scattering states take the form

$$\phi_{j\ell} = e^{iKX} \sin(k|x| + \theta_k) Y(|x|), \tag{6.143}$$

where $X = (j + \ell)/2, |x| = j - \ell, K = k_1 + k_2, k = (k_1 - k_2)/2$, and θ_k is the scattering phase shift, while $Y(|x|)$ is the Heaviside step function being zero at $x = 0$. A surprise in the two-body problem—which is general for lattice systems—comes when one introduces equation (6.143) into equation (6.142): the Hamiltonian depends parametrically on the total momentum K. It is easy to see, by writing $\phi_{j\ell} = \exp(iKX)\Phi(x)$ that the relative wave function $\Phi(x) = \Phi_K(x)$ and satisfies

$$\frac{J_K \hbar^2}{2}[\Phi_K(x + 1) + \Phi_K(x - 1)] + J\hbar^2 \delta_{|x|,1} \Phi_K(x) = \Delta E_K \Phi(|x|), \tag{6.144}$$

where $J_K = 2J\cos(K/2)$ carries the parametric K-dependence of the Schrödinger equation. Equation (6.144) has already been solved (it is now a one-body problem), and the reader is asked to investigate the consequences of non-zero total momentum in one of the problems at the end of the chapter.

6.2.2 Two bosons in a harmonic trap

In this section, we consider two bosons interacting via s-wave, LO EFT interaction in three spatial dimensions, under the influence of an external harmonic potential.

This problem is of great interest in ultracold atomic physics, since it is the simplest building block for interacting states of trapped atomic ensembles. Here, we partially follow the article by Busch and co-workers [5].

Most experiments nowadays have in place an optical lattice, or else an anharmonic potential. However, these are all harmonic around the potential minima, and modelling these—for two-body interacting energy shifts—as harmonic potentials may in fact be a very good approximation. The advantage of a harmonic potential is the fact that the centre-of-mass wave function is separable from the relative motion. Therefore, the problem simplifies enormously when compared to the anharmonic cases. Seeing this is straightforward. Let us consider the non-interacting Hamiltonian

$$H_0 = \frac{\mathbf{p}_1^2}{2m} + \frac{\mathbf{p}_2^2}{2m} + \frac{1}{2}m\omega^2(\mathbf{r}_1^2 + \mathbf{r}_2^2). \tag{6.145}$$

Defining centre-of-mass ($\mathbf{R}$) and relative ($\mathbf{r}$) coordinates in the usual manner, H_0 in equation (6.144) separates into $H_0 = H_0^{\mathrm{CM}} + H_0^{\mathrm{r}}$, with

$$H_0^{\mathrm{CM}} = \frac{\mathbf{P}^2}{2M} + \frac{1}{2}M\omega^2 R^2, \tag{6.146}$$

$$H_0^{\mathrm{r}} = \frac{\mathbf{p}^2}{2\mu} + \frac{1}{2}\mu\omega^2 R^2, \tag{6.147}$$

where $\mathbf{P}$ and $\mathbf{p}$ are, respectively, the centre-of-mass and relative momenta, $M = 2m$ is the total mass and $\mu = m/2$ is the reduced mass. Therefore, two interacting particles have a relative Hamiltonian given by

$$H^r = \frac{\mathbf{p}^2}{2\mu} + \frac{1}{2}\mu\omega^2 R^2 + V_{\mathrm{LO}}(\mathbf{r}). \tag{6.148}$$

We now solve the stationary Schrödinger equation for the relative Hamiltonian, equation (6.148), $H^r\psi = E\psi$. Since V_{LO} only acts on s-waves, we restrict ourselves to the channel with zero angular momentum. The wave function in spherical coordinates is given by $R(r)Y_{00}(\theta, \phi)$, where Y_{00} is the s-wave spherical harmonic, and the radial Schrödinger equation reads

$$-\frac{\hbar^2}{2\mu r^2}\frac{d}{dr}\left(r^2\frac{d}{dr}\right)R(r) + \frac{1}{2}\mu\omega^2 r^2 + \frac{g_R}{4\pi r^2}\delta(r)\frac{d}{dr}(rR(r)) = ER(r). \tag{6.149}$$

Above, for the LO interaction we have used the fact that $\delta(\mathbf{r}) = \delta(r)/4\pi r^2$. The solution to the Schrödinger equation (6.149) can be obtained analytically, and corresponds to a Gaussian times a confluent hypergeometric function, while the eigenenergies are obtained as

$$\sqrt{2}\frac{\Gamma(-\epsilon/2 + 3/4)}{\Gamma(-\epsilon/2 + 1/4)} = \frac{\ell}{a}, \tag{6.150}$$

where $\ell = \sqrt{\hbar/m\omega}$ is the oscillator length.

6.2.3 Two-body problem on a lattice

In section 6.1.4, we studied scattering of a particle off an impurity in a hypercubic lattice. Mostly, it is a solvable problem, although being a system with infinitely many channels, it is not trivial. In one spatial dimension, however, particle-impurity scattering is quite simple since the Green's functions can be calculated essentially analytically (given the non-interacting states).

The two-body problem on a realistic lattice is much more complicated than the particle-impurity collisions studied earlier. Although total quasi-momentum is conserved, the presence of the external potential makes the problem highly non-trivial, even in one spatial dimension. Writing down the Lippmann–Schwinger equation is straightforward (do it!), but its solution is not. What we can do, however, to alleviate this issue, at least at low energies, is to perform numerical diagonalization and use a generalization of Lüscher's method.

Let us particularize to the one-dimensional problem and, in particular, asymptotic behaviour of stationary scattering states. One may naïvely think that, for large interparticle separation $|z| = |z_1 - z_2| > R$, with R the range of the potential, the particles travel as non-interacting. That is, their asymptotic behaviour is simply given by the product of two Bloch waves. Unfortunately, the non-trivial, non-Galilean energy structure allows for the total quasi-momentum and total energy conservation in the two-body problem to be satisfied by more than one pair of quasi-momentum values. That is, if we have

$$E = \epsilon_{n_1}(k_1) + \epsilon_{n_2}(k_2), \tag{6.151}$$

$$K = k_1 + k_2 \pmod{2\pi/d}, \tag{6.152}$$

there is more than one non-trivial pair of quasi-momenta and band indices that satisfy the relations (6.151) and (6.152). This may in fact include complex pairs $k_1 = q_1 + i\lambda$ and $k_2 = K - q_1 - i\lambda$, with real q_1 and λ, provided their band indices are not equal to (n_1, n_2). Certain dispersions, moreover, may allow for several pairs of real quasi-momenta, and the same band indices, to satisfy the relations. This means there can be inelastic collisions, too. At low energies, however, where the dispersion $\epsilon_n(k)$ is approximately quadratic, inelastic collisions are impossible, and only complex quasi-momenta can satisfy the relations (6.151) and (6.152) along with the real quasi-momenta k_1 and k_2 we are interested in. Since complex quasi-momenta mean exponential tails for the extra wave functions, asymptotically and at low energies (and in the lowest band) we may safely assume that the particles move freely at long distances, up to exponentially small corrections.

Given the above analysis, in one spatial dimension, scattering at low energies can be easily analysed in finite boxes. To see this, take two bosons with a short-range interaction. The stationary states take the asymptotic form (as $|z_1 - z_2| \to \infty$)

$$\begin{aligned}\psi_{k_1,k_2}(z_1, z_2) \to\, & O_S\Big(e^{i(k_1z_1+k_2z_2)}\phi_{n_1,k_1}(z_1)\phi_{n_2,k_2}(z_2)\Big) \\ & + B\,\mathrm{sgn}(z_1 - z_2)O_A\Big(e^{i(k_1z_1+k_2z_2)}\phi_{n_1,k_1}(z_1)\phi_{n_2,k_2}(z_2)\Big),\end{aligned} \tag{6.153}$$

where O_S (O_A) is the symmetrization (antisymmetrization) operator[1], and B a constant that depends on (n_1, n_2) and (k_1, k_2), and contains the information on elastic scattering (i.e., the generalized phase shift). To see what this actually is, we only need to recall that, if no periodic potential is in place, then

$$B = -i \cot \theta_k. \tag{6.154}$$

We put the system in a finite box of size L with periodic boundary conditions. To make it simpler, L must be commensurate with the lattice spacing, that is, $L = Md$, with M a positive integer, and $L > d$. Up to exponential corrections, the quantization condition that allows us to extract B is obtained as follows: we may fix $z_2 = L/s$, with $s > 0$, in a way that the actual wave function is always asymptotic when we apply the boundary conditions for z_1. We then have

$$\psi_{k_1,k_2}(0, L/s) = \psi_{k_1,k_2}(L, L/s), \tag{6.155}$$

which gives

$$\begin{aligned} &O_S\left(e^{i(k_1L+k_2L/s)}\phi_{n_1,k_1}(L)\phi_{n_2,k_2}(L/s)\right) + BO_A\left(e^{i(k_1L+k_2L/s)}\phi_{n_1,k_1}(L)\phi_{n_2,k_2}(L/s)\right) = \\ &O_S\left(e^{ik_2L/s}\phi_{n_1,k_1}(0)\phi_{n_2,k_2}(L/s)\right) - BO_A\left(e^{ik_2L/s}\phi_{n_1,k_1}(0)\phi_{n_2,k_2}(L/s)\right), \end{aligned} \tag{6.156}$$

Now, since $\phi_{n,k}(0) = \phi_{n,k}(d)$ and $L = Md$, with M an integer, then $\phi_{n,k}(L) = \phi_{n,k}(0)$. If, moreover, s is an integer and M is divisible by s, then $\phi_{n,k}(L/s) = \phi_{n,k}(0)$. Considering this is the case, equation (6.156) simplifies to

$$\begin{aligned} &e^{i(k_1L+k_2L/s)} + e^{i(k_1L/s+k_2L)} - e^{ik_2L/s} - e^{ik_1L/s} = \\ &= B[e^{ik_1L/s} - e^{ik_2L/s} + e^{i(k_1L/s+k_2L)} - e^{i(k_1L+k_2L/s)}]. \end{aligned} \tag{6.157}$$

The most relevant case (but not general, since B does depend on $K = k_1 + k_2$!) corresponds to $K = 0$, which includes the ground state. In that case, $k_1 = -k_2 = k$, with k the relative momentum, and we have

$$\cos\left(\frac{k(s-1)L}{s}\right) - \cos(kL/s) = iB\left[\sin(kL/s) - \sin\left(\frac{k(s-1)L}{s}\right)\right]. \tag{6.158}$$

Obviously, the above relation is non-trivial only if we choose $1 - 1/s \neq 1/s$, that is, $s \neq 2$. The above relation is a quantization condition for k given B or, in our interpretation in which we know k (via the energy $E = 2\epsilon_0(k)$ in the case of $n_1 = n_2 = 0$), and we obtain B as

$$B = -i\frac{\cos\left(\frac{k(s-1)L}{s}\right) - \cos(kL/s)}{\sin(kL/s) - \sin\left(\frac{k(s-1)L}{s}\right)}, \tag{6.159}$$

[1] The convention used here is $O_{S,A}(f(x, y)) \equiv f(x, y) \pm f(y, x)$.

for any integer $s \neq 2$ such that M is divisible by s, and small enough that the relative wave function is always asymptotic in the boundary condition posted in equation (6.155). The reason why this can be done is because, in fact, the right-hand-side of equation (6.159) is s-independent, and can be simplified to

$$B = -i \tan\left(\frac{kL}{2}\right) \Rightarrow \cot \theta_k = \tan(kL/2), \tag{6.160}$$

which is exactly what one would obtain was there no lattice potential at all. This, in fact, is the major simplification we can obtain by choosing commensurate lengths.

We test now our prediction in equation (6.160) numerically, using the following model for the non-interacting Hamiltonian H_0 and the two-body interaction H_I,

$$H_0 = \sum_{i=1}^{2} \left[\frac{p_i^2}{2m} + V_0 \sin^2\left(\frac{2\pi z_i}{\tau}\right) \right], \tag{6.161}$$

$$H_I = g\delta(z_1 - z_2). \tag{6.162}$$

To solve the stationary Schrödinger equation numerically, we use the finite-difference method. We place the system on a grid with n points, equally spaced with a_0 the grid's spacing. The discrete Laplacian we use is given by

$$(\Delta f)(z) = \frac{1}{d^2}[f(z + a_0) + f(z - a_0) - 2f(z)]. \tag{6.163}$$

The Dirac delta function has a nice one-dimensional lattice approximation, which is simply

$$\delta(z) = \frac{\delta_{n_z,0}}{a_0}, \tag{6.164}$$

where $n_z a_0 = z$. We diagonalize the two-body discrete Hamiltonian to obtain the low-energy spectrum. In figure 6.5, we plot the energy shift, with respect to the non-interacting ground state, of two bosons on a lattice, which is not strong enough to be tight-binding but also not so weak that the effective mass approximation is valid for small system lengths. There, we observe that, as we expected, at commensurate lengths (half-integer values of L/τ), convergence is smooth as we lower the lattice spacing. Moreover, at commensurate lengths and large system sizes, the energy shifts approximately approach the free-space energy shifts. This is due to the effective mass approximation being valid at low quasi-momenta, which are dominant at large L. In the case we deal with in figure 6.5, the effective mass $m_*/m \approx 1.13$, which is quite close to 1 and the low-energy physics is similar to the continuum counterpart.

The phase shifts are obtained from equation (6.160), and the energy dispersion. We may displace the phase shift as $\delta_k = \theta_k + \pi/2$ so that equation (6.160) gives, for commensurate lengths,

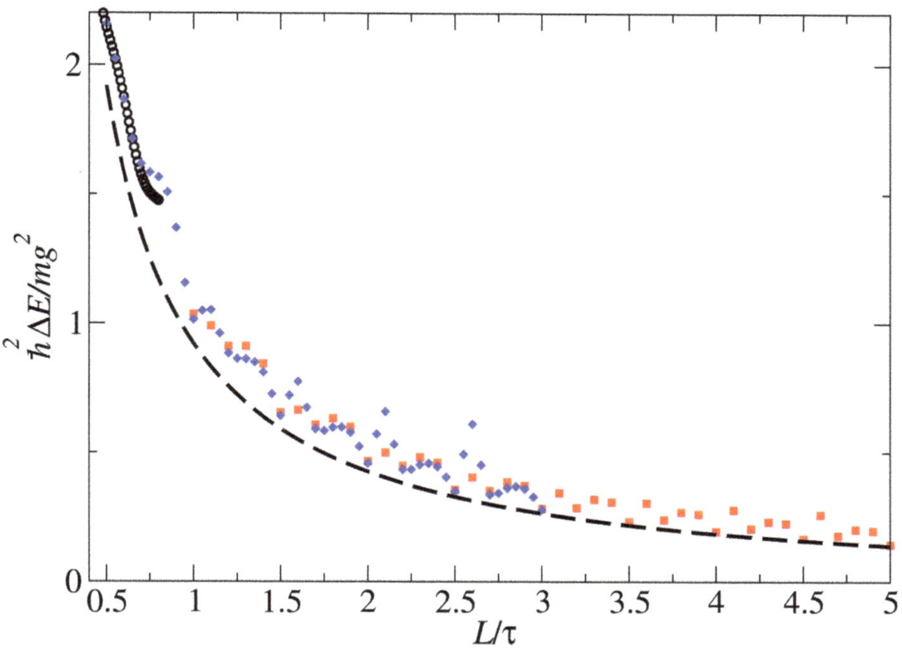

Figure 6.5. Black open circles, blue diamonds and red squares represent the energy shift, with increasing lattice spacing, as a function of systems length for $mg/\hbar^2\tau = 1$, for $\hbar^2 V_0/mg^2 = 40$. The black dashed line represents the free-space result ($V_0 \equiv 0$) with the same interaction strength.

$$\delta_k = -kL/2. \tag{6.165}$$

The scattering length a_* can be extracted easily for large L within the effective mass approximation. For small ka, we can approximate $\theta_k = -ka$, and therefore equation (6.166) gives

$$-ka + \frac{\pi}{2} \approx -\frac{kL}{2} \Rightarrow \Delta E = \frac{\hbar^2\pi^2}{m_*(L - 2a_*)^2}. \tag{6.166}$$

Using $m_* \approx 1.13m$, we obtain, from commensurate energy shifts, $a_*/a \approx 1.43$. Note that, although the estimate performed here is quite rough, one may obtain non-trivial scattering data out of the numerical diagonalization by using the full dispersion, and without the large L approximation. This method, we believe, is definitely much easier, at least at low energies, than solving the full two-body scattering problem on the lattice, and provides fully non-perturbative results.

Problems

1. Consider the one-dimensional scattering state of a particle, under the influence of a periodic potential $V(x)$, off a zero range impurity potential located at $x = 0$, given in equation (6.49). Calculate the transmission and reflection coefficients for a particle incoming from the left of the impurity.

2. Calculate the normalized eigenstates and associated eigenenergies in the infinite size limit of Hamiltonian (6.98).
3. Diagonalize numerically the following Hamiltonian:

$$H' = H + U\delta_{x,2},$$

where H is given in equation (6.98), in a box with open boundary conditions at $x = 0$ and $x = L$, as well as periodic boundary conditions in the y-direction. Explore the eigenenergies near the infinite size limit bound state, and show that the relation obtained from the naïve elastic scattering, equation (6.121), fails to describe what you observe.
4. Solve the Schrödinger equation (6.144), and show that for $K = \pi/d$, there is no motion in the relative coordinate. Show, moreover, that if a bound state exists, then it has infinite effective mass (this is called Mattis–Gallinar effect, see reference [6]).
5. Solve the one-dimensional scattering problem with LO odd-wave interactions for two spinless fermions. The bare interaction potential, in the momentum representation is given by

$$\langle k'|V|k\rangle = g_1 k'k.$$

References

[1] Watson G N 1939 Three triple integrals *Q. J. Math. Oxford Ser.* **2** 266
[2] Kogut J B 1979 An introduction to lattice gauge theory and spin systems *Rev. Mod. Phys.* **51** 659
[3] Bernevig B A and Hughes T L 2013 *Topological Insulators and Topological Superconductors* (Princeton, NJ: Princeton University Press)
[4] Qi X-L and Zhang S C 2011 Topological insulators and superconductors *Rev. Mod. Phys.* **83** 1057
[5] Busch T, Englert B G, Rzazewski K and Wilkens M 1998 Two particles in a harmonic trap *Found. Phys.* **28** 549
[6] Mattis D C and Gallinar J-P 1984 What is the mass of an exciton? *Phys. Rev. Lett.* **53** 1391

IOP Publishing

Strongly Interacting Quantum Systems, Volume 1
Few-body physics
Manuel Valiente and Nikolaj Thomas Zinner

Chapter 7

Three-body physics

So far, we have dealt with one- and two-body systems in a variety of settings. These are quite simple, at least *a priori*, since the equations needed to solve one- and two-particle scattering problems, as well as bound states, are usually easy enough to handle. The situation becomes much more intricate in the three-body problem, as we shall see in this chapter. Firstly, while two particles in free space can either scatter or bind together, three bodies have extra channels (for attractive interactions) corresponding to two bound particles scattering off a third. Moreover, the energies of these extra channels can be degenerate with those of three unbound particles, and the total momentum of a pair within the triplet is not conserved. Fortunately, we will see that getting our hands dirty with the three-body problem pays off, since there are a number of exotic phenomena associated with the quantum three-body problem, for both bound and scattering states. In this chapter, we introduce the formalism necessary to investigate the three-body problem and study in detail some of the most striking phenomenology known to date associated with, at least, the simplest interactions.

7.1 Faddeev equations

We begin by setting up the three-body equivalent of the fundamental equation of scattering theory, that is, the Lippmann–Schwinger equation for the transition operator, or T-matrix. As usual, we assume that the Hamiltonian can be written as $H = H_0 + V$, where H_0 is the total kinetic energy of the three-body system, while V is the total interaction. In most applications, one assumes that V only includes pairwise interactions, that is, interactions between pairs (12), (13) and (23). We will not assume this in the following, but instead consider as well a three-body interaction U, so that the total interaction takes the form

$$V = V^1 + V^2 + V^3 + U, \tag{7.1}$$

doi:10.1088/978-0-7503-3087-9ch7

where the superscript i ($i = 1, 2, 3$) indicates that the interaction V^i only involves particles $j \neq i$ and $\ell \neq i$. To clarify this point, in the position representation, a local interaction V is written as

$$V(\mathbf{r}_1, \mathbf{r}_2, \mathbf{r}_3) = V^1(\mathbf{r}_2, \mathbf{r}_3) + V^2(\mathbf{r}_1, \mathbf{r}_3) + V^3(\mathbf{r}_1, \mathbf{r}_2) + U(\mathbf{r}_1, \mathbf{r}_2, \mathbf{r}_3). \tag{7.2}$$

An alternative notation for V^i can be chosen, too, using the interacting pair labels as subscripts, i.e., $V^i = V_{j\ell}$. In this way, superscripts indicate the *spectator* particle(s), and subscripts indicate the *active* particles.

The momentum representation of the interaction is more transparent than its position counterpart (7.2) regarding conservation laws and other issues that will show up in an instant. Clearly, a two-body interaction V^i (out of the three possible) conserves both the total momentum of the pair $(j\ell)$, which we denote $\mathbf{K}^i \equiv \mathbf{K}_{j\ell}$, as well as the momentum of the spectator particle, which we denote $\mathbf{k}_i \equiv \mathbf{k}^{j\ell}$. Mathematically, we obviously write

$$\langle \mathbf{k}_1'\mathbf{k}_2'\mathbf{k}_3' | V^i | \mathbf{k}_1\mathbf{k}_2\mathbf{k}_3 \rangle = (2\pi)^{2D}\delta(\mathbf{k}_i - \mathbf{k}_i')\delta(\mathbf{K}^i - \mathbf{K}'^i)V^i(\mathbf{q}_{j\ell}), \tag{7.3}$$

where $\mathbf{q}_{j\ell} \equiv \mathbf{k}_{j\ell} - \mathbf{k}'_{j\ell}$, with $\mathbf{k}_{jl} = (\mathbf{k}_j - \mathbf{k}_\ell)/2$, and analogously for $\mathbf{k}'_{j\ell}$, denotes relative momentum. Note that, obviously, conservation of $\mathbf{k}_i$ and $\mathbf{K}^i$ implies conservation of total momentum $\mathbf{K} = \mathbf{K}^i + \mathbf{k}_i$. The genuine three-body interaction U is simpler, since only total momentum is conserved. We write

$$\langle \mathbf{k}_1'\mathbf{k}_2'\mathbf{k}_3' | U | \mathbf{k}_1\mathbf{k}_2\mathbf{k}_3 \rangle = (2\pi)^{D}\delta(\mathbf{K}^i - \mathbf{K}'^i)U(\mathbf{q}_{ij}, \mathbf{q}_{j\ell}). \tag{7.4}$$

The same notation used above for active and spectator particles is taken over to the Green's functions of the system. The non-interacting Green's function of the three-body set at (complex) energy z is denoted $G_0(z)$, and is given by $G_0(z) = (z - H_0)^{-1}$. This is the only non-interacting Green's function we need. As for interacting Green's functions, we must be a little observant in order not to make rookie mistakes. Denoting $H_{0,i} \equiv H_0^{j\ell}$ the non-interacting Hamiltonian for particle i, and $H_0^i = H_{0,j} + H_{0,\ell}$ the non-interacting Hamiltonian for the active particles j and ℓ, we define the interacting Green's function of the active pair $(j\ell)$, $G^i(z) \equiv G_{j\ell}(z)$ as

$$G^i(z) = (z - H_0^i - V^i)^{-1}. \tag{7.5}$$

We insist on the fact that $G^i(z)$ is an operator defined on the Hilbert space of the three-body system, even though it only depends explicitly on two-body operators. The total interacting Green's function is simply written $G(z) = (z - H)^{-1}$, while the Green's function associated with the genuine three-body potential U is denoted $G_U(z) = (z - H_0 - U)^{-1}$.

Everything is in place now to define our fundamental objects, that is, transition operators (T-matrices). We begin by defining the *disconnected* T-matrix for the pair $(j\ell)$, which we call $t^i(z)$, and the (connected) T-matrix for the three-body interaction, which we call $T_U(z)$. From the usual definition (see equation (1.69)) of the T-matrix, we have

$$t^i(z) = V^i + V^i G^i(z) V^i = V^i + V^i G_0(z) t^i(z), \tag{7.6}$$

$$T_U(z) = U + U G_U(z) U = U + U G_0(z) T_U(z). \tag{7.7}$$

The transition operator for the full three-body system, which we simply call $T(z)$, formally satisfies the Lippmann–Schwinger equation (1.73),

$$T(z) = V + V G_0(z) T(z). \tag{7.8}$$

The equation, as is, suffers from severe problems as far as its direct solution is concerned. The concrete mathematical issues are beyond the scope of this book, and can be consulted in the excellent classic textbook by C J Joachain [1]. These problems have to do with the connectivity of the T-matrix, and show up as Dirac delta functions that seem difficult to deal with.

To solve the aforementioned issues, we perform what is known as Faddeev decomposition of the transition operator. This consists of writing

$$T(z) = T_1(z) + T_2(z) + T_3(z) + W(z), \tag{7.9}$$

and inserting it into the Lippmann–Schwinger equation (7.8). We trivially obtain

$$\sum_{i=1}^{3} T_i(z) + W(z) = \sum_{i=1}^{3} V^i + U + \left(\sum_{i=1}^{3} V^i + U \right) G_0(z) \left[\sum_{i=1}^{3} T_i(z) + W(z) \right]. \tag{7.10}$$

We now separate the right-hand-side of equation (7.10) into four portions, each of which only contains one of the potential operators (one of $\{\{V^i\}_{i=1}^3, U\}$), so that we can split (7.10) into four equations

$$T_i(z) = V^i + V^i G_0(z) T_i(z) + V^i G_0(z) \left[T_j(z) + T_\ell(z) + W(z) \right] \quad (i = 1, 2, 3), \tag{7.11}$$

$$W(z) = U + U G_0(z) W(z) + U G_0(z) \sum_{j=1}^{3} T_j(z). \tag{7.12}$$

Observe that addition of the three equations (7.11), and equation (7.12) results in the original Lippmann–Schwinger equation, equation (7.8), as should be. We rearrange equation (7.11) as

$$(1 - V^i G_0(z)) T_i(z) = V^i + V^i G_0(z) \left[T_j(z) + T_\ell(z) + W(z) \right], \tag{7.13}$$

and analogously for equation (7.12). Multiplying equation (7.13) by $(1 - V^i G_0(z))^{-1}$ from the left, we recognize the disconnected T-matrix $t^i(z)$, equation (7.6), as

$$t^i(z) = (1 - V^i G_0(z))^{-1} V^i, \tag{7.14}$$

and analogously for T_U, equation (7.7), finally obtaining the so-called Faddeev equations

$$T_i(z) = t^i(z) + t^i(z) G_0(z) \left[T_j(z) + T_\ell(z) + W(z) \right] \quad (i = 1, 2, 3), \tag{7.15}$$

$$W(z) = T_U(z) + T_U(z)G_0(z)\left[T_i(z) + T_j(z) + T_\ell(z)\right]. \tag{7.16}$$

The observant reader will realize that equation (7.15) still has issues with connectivity. These are solved easily by iterating the Faddeev equations once. Simply insert equation (7.15) for T_j and T_ℓ into equation (7.15) for T_i, finally obtaining (we omit the z-dependence to avoid overloading the notation)

$$T_i = t^i + t^iG_0(t^j + t^\ell) + t^iG_0\left[t^jG_0(T_i + T_\ell + W) + t^\ell G_0(T_i + T_j + W) + W\right] \tag{7.17}$$

Note that all of the above equations can be written in matrix form, and the reader is encouraged to show this (see problem 1).

If we are interested in three-body bound states, the situation is slightly simpler. If there is a three-body bound state, then the full T-matrix has a pole at the bound state energy $z = E < E_2$, where E_2 is the lowest ground state energy among the three different two-body subsystems (12), (13), (23). By construction, the disconnected two-body T-matrices, equation (7.6), do not have poles, and writing $T_i(z) \equiv M_i(z)/(z - E)$, as well as $W(z) = N(z)/(z - E)$, and inserting these into equations (7.15) and (7.16), we have

$$M_i(z) = t^i(z)G_0(z)\left[M_j(z) + M_\ell(z) + N(z)\right], \tag{7.18}$$

$$N(z) = T_U(z)G_0(z)\left[M_i(z) + M_j(z) + M_\ell(z)\right]. \tag{7.19}$$

Inserting equation (7.19) into equation (7.18), we obtain a closed set of equations for M_i,

$$\begin{aligned} M_i(z) = {} & t^i(z)G_0(z)\left[M_j(z) + M_\ell(z)\right] \\ & + t^i(z)G_0(z)T_U(z)G_0(z)\left[M_i(z) + M_j(z) + M_\ell(z)\right]. \end{aligned} \tag{7.20}$$

These are the Faddeev equations for the bound states, which exist as long as equation (7.20) ($i = 1, 2, 3$) has negative energy solutions at $z = E < E_2$. Needless to say, the Faddeev equations are far more manageable when searching for bound rather than for scattering states. As we shall see, many of the interesting phenomena occur in the bound state sector, but we will also tackle some interesting questions in the collisional channels.

7.1.1 Bosonic bound states: Skornyakov–Ter-Martirosyan equation

Now that we have developed the Faddeev formalism for three particles, it is time to apply it to a very relevant problem in current research, namely three-body bound states of identical bosons in three spatial dimensions ($D = 3$), although we shall set up the final equation in any dimension. To make it as simple as possible, we will assume that the two-body s-wave scattering length is very large when compared to all other length scales in the problem, so that the two-boson interactions can be modelled by the leading-order (LO) EFT interaction (see chapter 3). Remarkably, while the two-body problem is easily renormalizable, as we have seen already, for

$D = 3$ the three-body problem requires further renormalization. That is, not only the two-body scale (scattering length), but also a three-body scale is necessary in order to obtain predictive power.

The bare LO interaction in the two-body subspace takes the usual constant form,

$$\langle \mathbf{k}' | V_{\text{LO}} | \mathbf{k} \rangle = g_0. \tag{7.21}$$

Since we do not have a three-body interaction to begin with, the Faddeev equations for the three-body bound states, equation (7.18), simplify to

$$M_i(z) = t^i(z) G_0(z) \big[M_j(z) + M_\ell(z) \big]. \tag{7.22}$$

We will work in the momentum representation. Therefore, it is necessary to begin with the momentum space structure of the disconnected two-body T-matrix $t^i(z)$. To simplify the notation, given a triple of vector momenta $(\mathbf{k}_1, \mathbf{k}_2, \mathbf{k}_3) \equiv (\mathbf{k}_i, \mathbf{k}_j, \mathbf{k}_\ell)$, we introduce the following notation

$$\vec{\mathbf{k}} \equiv (\mathbf{k}_i, \mathbf{k}_j, \mathbf{k}_\ell). \tag{7.23}$$

In this way, we can also write $|\vec{\mathbf{k}}\rangle \equiv |\mathbf{k}_i \mathbf{k}_j \mathbf{k}_\ell\rangle$ for a three-body momentum eigenstate. The T-matrix $t^i(z)$ has the same structure as the two-body potential *within the Hilbert space of the three-body system*. Therefore, from equation (7.3), we can write

$$\langle \vec{\mathbf{k}}' | t^i(z) | \vec{\mathbf{k}} \rangle = (2\pi)^{2D} \delta(\mathbf{k}_i - \mathbf{k}_i') \delta(\mathbf{K}^i - \mathbf{K}'^i) F_i(z; \mathbf{k}_{j\ell}, \mathbf{k}'_{j\ell}), \tag{7.24}$$

where F_i is some function of the energy z and the relative momenta $\mathbf{k}_{j\ell} = (\mathbf{k}_j - \mathbf{k}_\ell)/2$ and $\mathbf{k}'_{j\ell} = (\mathbf{k}'_j - \mathbf{k}'_\ell)/2$, as well as $\mathbf{k}_i$ and $\mathbf{K}^i$ (which is noted implicitly as a subscript). To see what the function F corresponds to, we write down equation (7.6) in the momentum representation. After easily getting rid of the Dirac delta functions, we have

$$F_i(z; \mathbf{k}_{j\ell}, \mathbf{k}'_{j\ell}) = \langle \mathbf{k}'_{j\ell} | V_2^i | \mathbf{k}_{j\ell} \rangle + \int \frac{d\mathbf{q}_{j\ell}}{(2\pi)^D} \frac{\langle \mathbf{k}'_{j\ell} | V_2^i | \mathbf{q}_{j\ell} \rangle}{z - \mathcal{E}_i(\mathbf{q}_{j\ell})} F_i(z; \mathbf{k}_{j\ell}, \mathbf{q}_{j\ell}). \tag{7.25}$$

Above, $\mathcal{E}_i$ is given by the three-body kinetic energy with conserved $\mathbf{k}_i$ and $\mathbf{K}^i$ and relative momentum $\mathbf{q}_{j\ell}$ between particles j and ℓ, namely

$$\mathcal{E}_i(\mathbf{q}_{j\ell}) = \epsilon_i(\mathbf{k}_i) + \epsilon_j(\mathbf{K}^i/2 + \mathbf{q}_{j\ell}) + \epsilon_\ell(\mathbf{K}^i/2 - \mathbf{q}_{j\ell}). \tag{7.26}$$

The resemblance of equation (7.25) for F_i with the usual two-body Lippmann–Schwinger in the momentum representation is obvious. The only difference lies in the fact that the kinetic energy function entering the non-interacting Green's function includes a contribution from all three particles. In the Galilean case, with $\epsilon_i(\mathbf{k}) = \hbar^2 k^2 / 2m_i$, we can safely neglect any total centre-of-mass effects and set the total momentum $\mathbf{K} = \mathbf{k}_i + \mathbf{K}^i = 0$. Since we are dealing with bosons, then $m_i = m_j = m_\ell \equiv m$. The kinetic energy (7.26) simplifies to

$$\mathcal{E}_i(\mathbf{q}_{j\ell}) = \frac{\hbar^2 q_{j\ell}^2}{m} + \frac{3\hbar^2}{4m}k_i^2. \tag{7.27}$$

Inserting this into equation (7.25), simple comparison with the Lippmann–Schwinger equation satisfied by the two-body T-matrix $\mathcal{T}(z)$ in the two-particle subspace, implies that in the Galilean case, we have

$$F_i(z; \mathbf{k}_{j\ell}, \mathbf{k}'_{j\ell}) = \langle \mathbf{k}'_{j\ell}| \mathcal{T}\left(z - \frac{3\hbar^2}{4m}k_i^2\right)|\mathbf{k}_{j\ell}\rangle. \tag{7.28}$$

This is the first instance of an off-shell T-matrix appearing in the three-body problem. This, in fact, is recurrent for more than two particles, where the off-shell structure of the two-body T-matrices contributes to higher-particle-number processes.

We derive now the so-called Skornyakov–Ter-Martirosyan equation, in its bare (non-renormalized) form first. We set the total momentum to $\mathbf{K} = 0$, and allow ourselves to have a rather inconsequential abuse of notation, by forcing $\mathbf{K} = \mathbf{k}_i + \mathbf{k}_j + \mathbf{k}_\ell = 0$ below and setting

$$\langle \mathbf{k}'|M_i(z)|\mathbf{k}\rangle = (2\pi)^{2D}\delta(\mathbf{K}')\langle \vec{\mathbf{k}}'|M_i(z)|\vec{\mathbf{k}}\rangle. \tag{7.29}$$

On the right-hand-side of the relation above, it is implicitly assumed that $\mathbf{k}_\ell = -\mathbf{k}_i - \mathbf{k}_j$, and analogously for $\mathbf{k}'_\ell$. Analogous notation is used for the matrix elements of $t^i(z)$. With this in mind, we are ready to write equation (7.22) for the bound states in the momentum representation. This is a standard exercise, and we have

$$\begin{aligned}\langle \vec{\mathbf{k}}'|M_i(z)|\vec{\mathbf{k}}\rangle = \int \frac{d\mathbf{q}_i}{(2\pi)^D}\int \frac{d\mathbf{q}_j}{(2\pi)^D}\frac{\langle \mathbf{k}'_i\mathbf{k}'_j, -\mathbf{k}'_i - \mathbf{k}'_j|t^i(z)|\mathbf{q}_i\mathbf{q}_j, -\mathbf{q}_i - \mathbf{q}_j\rangle}{z - \epsilon(\mathbf{q}_i) - \epsilon(\mathbf{q}_j) - \epsilon(\mathbf{q}_i + \mathbf{q}_j)} \\ \times \langle \mathbf{q}_i\mathbf{q}_j, -\mathbf{q}_i - \mathbf{q}_j|M_i(z) + M_j(z)|\mathbf{k}_i\mathbf{k}_j, -\mathbf{k}_j - \mathbf{k}_\ell\rangle.\end{aligned} \tag{7.30}$$

Inserting equations (7.24) and (7.28) into equation (7.30), this simplifies to

$$\langle \vec{\mathbf{k}}'|M_i(z)|\vec{\mathbf{k}}\rangle = \mathcal{T}\left(z - \frac{3\hbar^2 k_i'^2}{4m}\right)\int \frac{d\mathbf{q}_j}{(2\pi)^D}\frac{\langle \mathbf{k}'_i\mathbf{q}_j, -\mathbf{k}'_i - \mathbf{q}_j|M_i(z) + M_j(z)|\mathbf{k}_i\mathbf{k}_j, -\mathbf{k}_j - \mathbf{k}_\ell\rangle}{z - \epsilon(\mathbf{k}'_i) - \epsilon(\mathbf{q}_j) - \epsilon(\mathbf{k}'_i + \mathbf{q}_j)}. \tag{7.31}$$

Clearly, we see from the equation above that $\langle \vec{\mathbf{k}}'|M_i(z)|\vec{\mathbf{k}}\rangle$ only depends on $\mathbf{k}'_i$, and we can write

$$\langle \vec{\mathbf{k}}'|M_i(z)|\vec{\mathbf{k}}\rangle \equiv M_i(z; \mathbf{k}'_i). \tag{7.32}$$

Writing the corresponding integral equations for M_j and M_ℓ (analogous to equation (7.32)), we see that the matrix elements of M_j and M_ℓ depend solely, respectively, on $\mathbf{k}'_j$ and $\mathbf{k}'_\ell$. And because of bosonic symmetry, their functional forms are all identical, that is

$$\langle \vec{\mathbf{k}}'|M_i(z)|\vec{\mathbf{k}}\rangle \equiv M_i(z; \mathbf{k}'_i) \equiv M(z; \mathbf{k}'_i), \tag{7.33}$$

$$\langle\vec{\mathbf{k}}'|M_j(z)|\vec{\mathbf{k}}\rangle = M(z;\, \mathbf{k}'_j), \tag{7.34}$$

$$\langle\vec{\mathbf{k}}'|M_\ell(z)|\vec{\mathbf{k}}\rangle = M(z;\, \mathbf{k}'_\ell). \tag{7.35}$$

Inserting equations (7.33)–(7.35) into equation (7.32), and assigning new labels to the dummy variables as $\mathbf{k}'_i = \mathbf{k}$, $\mathbf{q}_j = \mathbf{q}$, we finally obtain the following equation for the bound states of three non-relativistic bosons at vanishing total momentum

$$M(z;\, \mathbf{k}) = \mathcal{T}\left(z - \frac{3\hbar^2k^2}{4m}\right)\int \frac{d\mathbf{q}}{(2\pi)^D}\frac{M(z;\, \mathbf{q}) + M(z;\, -\mathbf{k} - \mathbf{q})}{z - \epsilon(\mathbf{k}) - \epsilon(\mathbf{q}) - \epsilon(\mathbf{k} + \mathbf{q})}. \tag{7.36}$$

The second term can be changed due to symmetry, under the integral sign, as $M(z;\, -\mathbf{k} - \mathbf{q}) \to M(z;\, \mathbf{q})$, and we arrive at the bare Skornyakov–Ter-Martirosyan equation

$$M(z;\, \mathbf{k}) = 2\mathcal{T}\left(z - \frac{3\hbar^2k^2}{4m}\right)\int \frac{d\mathbf{q}}{(2\pi)^D}\frac{M(z;\, \mathbf{q})}{z - \epsilon(\mathbf{k}) - \epsilon(\mathbf{q}) - \epsilon(\mathbf{k} + \mathbf{q})}. \tag{7.37}$$

Since we are searching for bound states, we also have $z = E = -|E| < 0$. The interaction, in this case, is the leading-order s-wave potential, given in its bare form by equation (7.21). Upon renormalization, the two-body T-matrix in $D = 3$ is given by (see chapter 3),

$$\mathcal{T}(-|E| = -\hbar^2Q^2/m) = \frac{4\pi\hbar^2}{m}\frac{1}{1/a - Q}, \tag{7.38}$$

which gives problems for three particles for $D = 3$. We shall solve these in the next subsection, and demonstrate the Efimov effect.

7.1.2 Thomas collapse and the Efimov effect

We consider the three-dimensional case (3D) of three identical bosons with LO s-wave interactions. In particular, our goal will be to describe the so-called Efimov effect: an infinite series of three-body bound states (at unitarity) whose energies satisfy a discrete scaling relation. Before showing the Efimov effect fully analytically, we show that, without renormalization of the three-boson problem beyond the usual two-body scattering amplitude, the system exhibits unphysical behaviour, in the form of bound states without a lower energy bound. This is called Thomas collapse. The Efimov effect is a prime example, with experimental relevance, of the breaking of continuous scale invariance down to its discrete counterpart, and of an exactly solvable three-body problem.

Our starting point is the Skornyakov–Ter-Martirosyan equation, equation (7.37), for $D = 3$. Because of rotational symmetry, we can work directly in the s-wave channel, for which M only depends on $k = |\mathbf{k}|$. Setting $z = -|E| = -\hbar^2Q^2/m$, and integrating over the solid angle yields the following one-dimensional integral equation

$$M(k) = 2\mathcal{T}\left(-\frac{\hbar^2}{m}(Q^2 + 3k^2/4)\right)\frac{m}{\hbar^2 k}\int_0^\infty \frac{dq}{(2\pi)^2}\ln\left|\frac{Q^2 + k^2 + q^2 - kq}{Q^2 + k^2 + q^2 + kq}\right| qM(q). \quad (7.39)$$

Above, we have removed the explicit energy dependence of M for simplicity, and the infinite upper limit of integration is to be substituted with a momentum cutoff Λ should it be required. In equation (7.39), the kernel is symmetric, i.e.,

$$\bar{\mathcal{K}}(k, q) = q\ln\left|\frac{Q^2 + k^2 + q^2 - kq}{Q^2 + k^2 + q^2 + kq}\right| = \bar{\mathcal{K}}(k, -q). \quad (7.40)$$

If we construct a symmetric extension $\tilde{M}$ of M to the whole real line $(-\infty, \infty)$, we can divide equation (7.39) by 2 and extend the integration domain to the real line. Further defining $f(k) \equiv k\tilde{M}(k)$, we have

$$f(k) = \mathcal{T}\left(-\frac{\hbar^2}{m}(Q^2 + 3k^2/4)\right)\frac{m}{\hbar^2}\int_{-\infty}^\infty \frac{dq}{(2\pi)^2}\mathcal{K}(k, q)f(q), \quad (7.41)$$

where $\mathcal{K}(k, q) = \bar{\mathcal{K}}(k, q)/q$. The above integral equation, equation (7.41), is, quite remarkably, exactly solvable. We follow the solution given by Gogolin, Mora and Egger in reference [2]. We begin by performing a change of variables

$$k = \frac{2}{\sqrt{3}}Q\sinh\xi, \quad (7.42)$$

$$q = \frac{2}{\sqrt{3}}Q\sinh\xi'. \quad (7.43)$$

We define $g(\xi) \equiv f(2Q\sinh\xi/\sqrt{3})$, and equation (7.41) transforms into

$$\begin{aligned} g(\xi) = \mathcal{T}\left(-\frac{\hbar^2}{m}(Q^2 + 3k^2/4)\right)\frac{2mQ}{\sqrt{3}\hbar^2}\int_{-\infty}^\infty \frac{d\xi'}{(2\pi)^2}g(\xi')\cosh\xi' \\ \times\ln\left|\frac{-1 + e^{2\xi'} + e^{-2\xi'} + e^{2\xi} + e^{-2\xi} - e^{\xi+\xi'} + e^{\xi-\xi'} + e^{\xi'-\xi} - e^{-\xi-\xi'}}{-1 + e^{2\xi'} + e^{-2\xi'} + e^{2\xi} + e^{-2\xi} + e^{\xi+\xi'} - e^{\xi-\xi'} - e^{\xi'-\xi} + e^{-\xi-\xi'}}\right|. \end{aligned} \quad (7.44)$$

The argument in the logarithm can be simplified significantly. It is easy to show that the numerator equals

$$(e^{2\xi} - e^{\xi+\xi'} + e^{2\xi'})(e^{2(\xi+\xi')} + e^{\xi+\xi'} + 1)e^{-2(\xi+\xi')}, \quad (7.45)$$

while the denominator can be rewritten as

$$(e^{2\xi} + e^{\xi+\xi'} + e^{2\xi'})(e^{2(\xi+\xi')} - e^{\xi+\xi'} + 1)e^{-2(\xi+\xi')}. \quad (7.46)$$

Therefore, the logarithm in equation (7.44) can be split as

$$\ln\left|\frac{e^{2\xi} - e^{\xi+\xi'} + e^{2\xi'}}{e^{2\xi} + e^{\xi+\xi'} + e^{2\xi'}}\right| + \ln\left|\frac{e^{2(\xi+\xi')} + e^{\xi+\xi'} + 1}{e^{2(\xi+\xi')} - e^{\xi+\xi'} + 1}\right|. \quad (7.47)$$

Equation (7.44) is simplified by noticing that if we transform $\xi' \to -\xi'$, then the second term in equation (7.47) equals the first term with the opposite sign. Since the function $g(\xi') = -g(-\xi')$, both terms contribute equally to the integral on the right-hand-side of equation (7.44). Therefore, we obtain

$$g(\xi) = \mathcal{T}\left(-\frac{\hbar^2}{m}(Q^2 + 3k^2/4)\right)\frac{4mQ}{\sqrt{3}\hbar^2}\int_{-\infty}^{\infty}\frac{d\xi'}{(2\pi)^2}g(\xi')\cosh\xi' \ln\left|\frac{e^{2\xi} - e^{\xi+\xi'} + e^{2\xi'}}{e^{2\xi} + e^{\xi+\xi'} + e^{2\xi'}}\right|. \tag{7.48}$$

We insert the value of the two-body T-matrix, equation (7.37), into equation (7.48), together with the change of variable (7.42), and define $\cosh\xi g(\xi) = h(\xi)$. We obtain

$$\left[1 - \frac{1}{Qa\cosh\xi}\right]h(\xi) = \frac{4}{\sqrt{3}\pi}\int_{-\infty}^{\infty} d\xi' h(\xi')\ln\left|\frac{e^{2\xi} - e^{\xi+\xi'} + e^{2\xi'}}{e^{2\xi} + e^{\xi+\xi'} + e^{2\xi'}}\right|. \tag{7.49}$$

We particularize to the unitary limit ($1/a = 0$). We perform yet another change of variables, by setting $t = \exp(\xi)$ and $t' = \exp(\xi')$, and redefine the function $h(\xi) = \chi(t)$, obtaining

$$\chi(t) = \frac{4}{\sqrt{3}\pi}\int_0^{\infty} dt' t'^{-1}\chi(t')\ln\left|\frac{e^{2\xi} - e^{\xi+\xi'} + e^{2\xi'}}{e^{2\xi} + e^{\xi+\xi'} + e^{2\xi'}}\right|. \tag{7.50}$$

The above equation is scale invariant (as should be for infinite scattering length), and this allows us to search for power-law solutions to the equation. We set $\chi(t) = t^s$, and perform the change of variables $t'/t = z$. Then, equation (7.50) reduces to

$$1 = \frac{4}{\sqrt{3}\pi}\int_0^{\infty} dz z^{s-1}\ln\left|\frac{z^2 + z + 1}{z^2 - z + 1}\right|. \tag{7.51}$$

The integral above corresponds to a Mellin transform, tabulated in reference [3], obtaining

$$s = \frac{8}{\sqrt{3}}\frac{\sin(\pi s/6)}{\cos(\pi s/2)}, \tag{7.52}$$

with $-1 < \mathrm{Re}(s) < 1$. The roots to the above equation are complex and given by $s = \pm i s_0$, with $s_0 = 1.006\,237\,825\,102\,78\ldots$. Hence, the solution to our problem is a combination of t^{is_0} and t^{-is_0}, that is

$$\chi(t) = \alpha t^{is_0} + \beta t^{-is_0}. \tag{7.53}$$

To establish the relation between the coefficients α and β above, we should trace back to the original function $M(k)$. This should be regular at the origin of momentum, and therefore $f(k) = kM(k)$ must satisfy $f(0) = 0$, which implies $g(0) = f(0) = 0$, which in turn makes $\chi(1) = 0$. Therefore, from equation (7.53) we infer $\beta = -\alpha$, so that we can write

$$M(k) = \frac{1}{k}\frac{1}{\sqrt{Q^2 + 3k^2/4}}\sin\left(s_0 \operatorname{arcsinh}\left(\frac{\sqrt{3}}{2}\frac{k}{Q}\right)\right). \tag{7.54}$$

The solution $M(k)$ to the Skornyakov–Ter-Martirosyan equation is valid for the continuum of positive values of Q, and therefore there is a bound state solution for all negative energies $E(Q) = -\hbar^2 Q^2/m$. The spectrum is unbound from below, and every energy is allowed, corresponding to an (unphysical!) effect known as Thomas collapse. Without entering the rigorous mathematical reasons why this is the case, we can easily see why, albeit well defined, the solutions $M(k)$ in equation (7.55) are unphysical. Take the high-momentum limit of equation (7.55), by expanding the arcsinh function at large momenta. We have

$$M(k) \propto \frac{1}{k^2} \sin\left(\frac{s_0}{2} \ln\left(\frac{3k^2}{Q^2}\right)\right), \qquad k \to \infty. \tag{7.55}$$

The problem with equation (7.55) is the unavoidable dependence of the large-k limit of $M(k)$, and therefore the wave function, with the binding energy ($\propto Q^2$). This is unphysical, since it implies that the short-range asymptotics of the wave function depend on the energy. This is quite simple to prove, and we shall do so in the second volume of this book.

The above analysis has shown us how a perfectly well defined, renormalizable EFT at the two-body level—in fact, the simplest one—fails to describe three-boson physics correctly. One may think that the unitarity limit is special and that the problem might be solved by working with finite scattering lengths. This is unfortunately not the case. What we do know is that the problem we have thus far tackled, in the unitary limit, exhibits (unbroken) continuous scale invariance. This is the reason why, if there is a bound state at energy E, then there is a bound state at energy $E' = \lambda E$, for *every* $\lambda > 0$. Since the integral equation has a solution at some negative energy, it must have a solution at every negative energy, and therefore the Thomas collapse is nothing but the unavoidable consequence of scale invariance and the existence of a bound state. If a physical three-boson bound state is to exist, then we must break scale invariance, that is, we need to introduce a further length scale besides the scattering length. This will be done with a (inverse) length scale, a so-called (non-unique) three-body parameter, which will fix the problem via renormalization. There are many ways to do this, and we explain here a few of these.

The first point we can make here is due to Danilov [4]. The short-distance behaviour (large k) of the wave functions is incorrect, as we have already mentioned. All wave functions should have a *weak universal* form at large momenta [5]. Hence, to try and fix the issues, we choose a fixed $Q = \Lambda_*$ and impose that the solutions must behave as

$$M(k) \propto \frac{1}{k^2} \sin\left(\frac{s_0}{2} \ln\left(\frac{3k^2}{\Lambda_*^2}\right)\right), \qquad k \to \infty. \tag{7.56}$$

In this case, Λ_* would be the so-called three-body parameter. Equation (7.56) has discrete scale invariance, since it remains unchanged upon the transformation

$$\Lambda_* \to \Lambda_* e^{n\pi/s_0}, \qquad n \in \mathbb{Z}. \tag{7.57}$$

Regarding Λ_* as a binding momentum, then we have an infinite, discrete set of allowed binding momenta $Q_n = \Lambda_* \exp(n\pi/s_0)$. Unfortunately, this still implies that the spectrum is unbound from below. We need a procedure that sets the ground state energy of the system to an arbitrary, but fixed and finite negative value.

Method I. Standard renormalization. By this point, the reader should know how to renormalize an operator (see chapter 3). So let us begin with this method by writing down the Lippmann–Schwinger equation for the full, bare, three-body T-matrix once more

$$T(z) = V + VG_0(z)T(z), \tag{7.58}$$

where V is the bare interaction (which contains three terms, as usual). We assume we know the T-matrix at some particular energy scale μ, and we obtain, using equation (3.116),

$$T(z) = T(\mu) + T(\mu)[G_0(z) - G_0(\mu)]T(z). \tag{7.59}$$

As input, we need the full three-body T-matrix at a particular energy scale μ. Of course, if $\mu > 0$, it is understood that $\mu \to \mu + i\eta$ ($\eta \to 0^+$). The structure of the three-body T-matrix is given by the Faddeev decomposition, equation (7.9), where T_i and $W(z)$ are solutions to the Faddeev equations (7.16) and (7.17). Therefore, we must follow this structure when setting the structure of the (operator-valued) renormalization condition, i.e., the value of $T(\mu)$.

We begin with a choice of renormalization condition that will lead to Danilov's breakdown of continuous scale invariance, but with the same issue regarding the spectrum being unbound from below. We remind the reader that we work at unitarity ($1/a = 0$). This choice consists of the following renormalization condition, with $\mu < 0$:

$$T(z \to \mu) = \frac{\tau(\mu)}{z - \mu}, \tag{7.60}$$

that is, we impose that there is a bound state with energy μ. Above, $\tau(\mu)$ is an operator that we must define for the renormalization condition in equation (7.60) to be complete. Note that equation (7.60) only specifies the dominant term near $z = \mu$, and is not the full T-matrix at all z, which can be written as

$$T(z) = \tau(z)/(z - \mu), \tag{7.61}$$

allowing $\tau(z)$ to become singular for $z \neq \mu$. That is, $\tau(\mu)$ does not specify the T-matrix everywhere, and there can be other poles other than $z = \mu$. We define $\tau(\mu)$ as

$$\tau(\mu) = M_i(\mu) + M_j(\mu) + M_\ell(\mu), \tag{7.62}$$

where M_i, M_j and M_ℓ have the same structure as those in equations (7.33–7.35), but are not necessarily the same solutions. In fact, due to scale invariance, the structure of the solution at the renormalization scale is sufficient to determine subsequent solutions at $z \neq \mu$. We insert equations (7.60) and (7.61) into equation (7.59), obtaining

$$\tau(\mu)[G_0(z) - G_0(\mu)]\tau(z) = -\tau(\mu). \quad (7.63)$$

We search for bound state solutions at energies $E \neq \mu$, and we can therefore write, for z near E

$$\tau(z \to E) = \frac{A(E)}{z - E}. \quad (7.64)$$

Inserting now equation (7.62) and equation (7.64) into equation (7.63), and using the structure of the functions M_i, M_j and M_ℓ in equations (7.33–7.35), we easily see that $A(E)$ satisfies

$$0 = \int \frac{d\mathbf{q}_i}{(2\pi)^3} \int \frac{d\mathbf{q}_j}{(2\pi)^3} [G_0(E; \vec{\mathbf{q}}) - G_0(\mu; \vec{\mathbf{q}})]A(E; \vec{\mathbf{q}}). \quad (7.65)$$

The notation in the above equation should be obvious by now. Equation (7.65) is rather simple-looking. In fact, we can try an Ansatz that exactly solves equation (7.65), namely

$$A(E, \vec{\mathbf{q}}) \equiv b(E; \mathbf{q}_i)a(E; \mathbf{q}_j). \quad (7.66)$$

For A to be a solution to equation (7.65), it suffices for a to satisfy

$$\int \frac{d\mathbf{q}_j}{(2\pi)^3} [G_0(E; \vec{\mathbf{q}}) - G_0(\mu; \vec{\mathbf{q}})]a(E; \mathbf{q}_j) = 0, \quad \forall \mathbf{q}_i. \quad (7.67)$$

It is a simple exercise (see problem 2), completely identical to the original Thomas collapse, to show that if the bound state energies are defined as $E = -\hbar^2 Q^2/m$, and $-\mu \equiv \hbar^2 \Lambda_*^2/m$, then the solutions are given by

$$Q = Q_n = \Lambda_* e^{n\pi/s_0}, \quad n \in \mathbb{Z}. \quad (7.68)$$

That is, we have managed, by introducing the scale μ, to break the scale invariance down to its discrete version, with the correct scaling factor, but the spectrum remains unbound from below. What has gone wrong here is the structure of the bound state at the renormalization scale. We shall therefore take a different route.

For our second (and good!) try, we will partially follow reference [6], with great pedagogical value. For this second choice, we assume that the full three-body T-matrix at the renormalization scale (and only there) is completely disconnected. That is, we assume that at $z = \mu$, we have

$$T(\mu) = t^i(\mu) + t^j(\mu) + t^\ell(\mu), \quad (7.69)$$

where t^i is the two-body T-matrix with spectator particle i. Note that in the original Faddeev equations, equation (7.69) is true for $\mu \to -\infty$, which has no scale, and equation (7.69) is putting in a scale at which the lowest-order approximation to the off-shell solution of the Faddeev equations becomes exact. We are working with the leading-order bosonic EFT. Renormalization condition (7.69) has the great advantage that the bound state equation reduces to the unrenormalized one, equation

(7.22), with $G_0(z)$ replaced with $G_0(z) - G_0(\mu)$. This is very easy to verify, and we urge the reader to do so. The kernel, equation (7.40), of the integral equation (7.41), in turn, becomes

$$\mathcal{K}(k, q) = \ln \left| \frac{Q^2 + k^2 + q^2 - kq}{Q^2 + k^2 + q^2 + kq} \right| - \ln \left| \frac{Q_\mu^2 + k^2 + q^2 - kq}{Q_\mu^2 + k^2 + q^2 + kq} \right|, \tag{7.70}$$

where the energy is $E = -\hbar^2 Q^2/m$ and the renormalization scale is $\mu = -\hbar^2 Q_\mu^2/m$. A numerical solution to the integral equation (7.41) with the renormalized kernel (7.71) shows that, indeed, the ground state energy is now finite, and that the excited state energies follow the correct Efimov scaling (see equation (7.68)), if we define the ground state energy $E_0 \equiv -\hbar^2\Lambda_*^2/m$.

Method II. Danilov–Pricoupenko filtering. There is an essentially equivalent method to the standard renormalization procedure to obtain a physically admissible bound state spectrum for three bosons at unitarity. This is similar in spirit to Danilov's original arguments, and is detailed in reference [7]. What one does is to fix a node for every bound state function $M(k)$, which we denote k_*. This in turn becomes an auxiliary condition, so that our solutions are now given by the integral equation (7.39) with the constraint $M(k_*) = 0$. This simply restricts which solutions to equation (7.39) are physically admissible.

The nodal condition can be easily analysed asymptotically. From equation (7.56), the nodal condition reads

$$\sin\left(\frac{s_0}{2} \ln\left(\frac{3k_*^2}{Q^2}\right)\right) = 0 \Rightarrow s_0 \ln|\sqrt{3}k_*/Q| = n\pi, \quad n \in \mathbb{Z}, \tag{7.71}$$

or, equivalently, the binding momenta $Q = Q_n$ satisfy

$$Q_n = \sqrt{3}k_* e^{-n\pi/s_0}, \quad n \in \mathbb{Z}. \tag{7.72}$$

With the above, it appears that we have not solved the issue of unboundedness of the spectrum from below. However, the reader should have noticed that, for equation (7.72) to be valid, the function $M(k)$ must be in the asymptotic region, that is, equation (7.72) is only valid for $Q_n \ll \sqrt{3}k_*$. If the possible binding momenta are large (of the order of k_* or larger), then the full expression for $M(k)$, equation (7.54), must be used. So let us assume that $Q = Q_n \gg k_*$. We can expand the arcsinh function in powers of k/Q, and, to leading-order, we find that the nodal condition is given by

$$s_0 \frac{\sqrt{3}}{2} \frac{k_*}{Q} = n'\pi, \quad n' \in \mathbb{Z}_+ - \{0\}, \tag{7.73}$$

where we have removed the possibility of $k_* = 0$ because necessarily $M(0) \neq 0$ (see equation (7.54)). Therefore, $Q/k_* = s_0\sqrt{3}/2n'\pi \approx 0.28/n'$, with n' a positive integer, together with the condition that $Q/k_* \gg 1$. These two conditions contradict each other, and therefore the spectrum must be bound from below. In fact, the ground state energy can be calculated from the first (smallest) positive node k_* as

$$s_0 \text{arcsinh}\left(\frac{\sqrt{3}}{2}\frac{k_*}{Q_0}\right) = \pi \Rightarrow Q_0 \approx 0.0765 k_*. \tag{7.74}$$

Therefore, we conclude that, within Danilov–Pricoupenko filtering, the binding momenta satisfy

$$Q_n \approx 0.0765 k_* e^{-n\pi/s_0}, \quad n = 0, 1, \ldots. \tag{7.75}$$

The above relation is the Efimov effect, now calculated analytically and with a well defined ground state.

To verify all of the above numerically, we go back to the Skornyakov–Ter-Martirosyan equation, equation (7.39). We insert the nodal condition, $M(k_*) = 0$, into the equation, with $k = k_*$, obtaining

$$\int_0^\infty dq \tilde{\mathcal{K}}(k_*, q) M(q) = 0. \tag{7.76}$$

Because of this, we may replace the kernel $\tilde{K}(k, q)$ by the subtracted kernel

$$\tilde{\mathcal{K}}_{k_*}(k, q) = \tilde{\mathcal{K}}(k, q) - \tilde{\mathcal{K}}(k_*, q). \tag{7.77}$$

Numerical results for the spectrum of the three-boson problem at unitarity using Danilov–Pricoupenko filtering are shown in figures 7.1 and 7.2.

Method III. Three-body counterterm. There is yet another method to obtain the correct spectrum of Efimov states, using a three-body counterterm in the Skornyakov–Ter-Martirosyan equation. This is detailed in reference [8]. The method, as featured in the original publication, seems rather sophisticated, and is obtained by summing infinitely many diagrams. This is quite unnecessary for a few-body problem of the simplicity we are dealing with here, as we shall see.

We begin by setting the system to unitarity (infinite two-body scattering length), and allowing for a genuine three-body interaction U to be present. This does not complicate the Faddeev equations in any significant way, as we have already seen. The Faddeev components M_i for general interactions obey equation (7.20). If the three-body interaction is a contact (and constant) potential, the Faddeev equations reduce to a single equation for M, given by equation (7.37) plus the following term added on its right-hand-side:

$$3\mathcal{T}\left(z - \frac{3}{4}\frac{\hbar^2 k^2}{m}\right) T_U(z) \int \frac{d\mathbf{q}}{(2\pi)^D} I(z; k) I(z; q) M(z; \mathbf{q}), \tag{7.78}$$

where $I(z; k)$ is an integral of the form

$$I(z; k) = \int \frac{d\mathbf{q}}{(2\pi)^D} \frac{1}{z - \epsilon(\mathbf{k}) - \epsilon(\mathbf{q}) - \epsilon(\mathbf{k} + \mathbf{q})}. \tag{7.79}$$

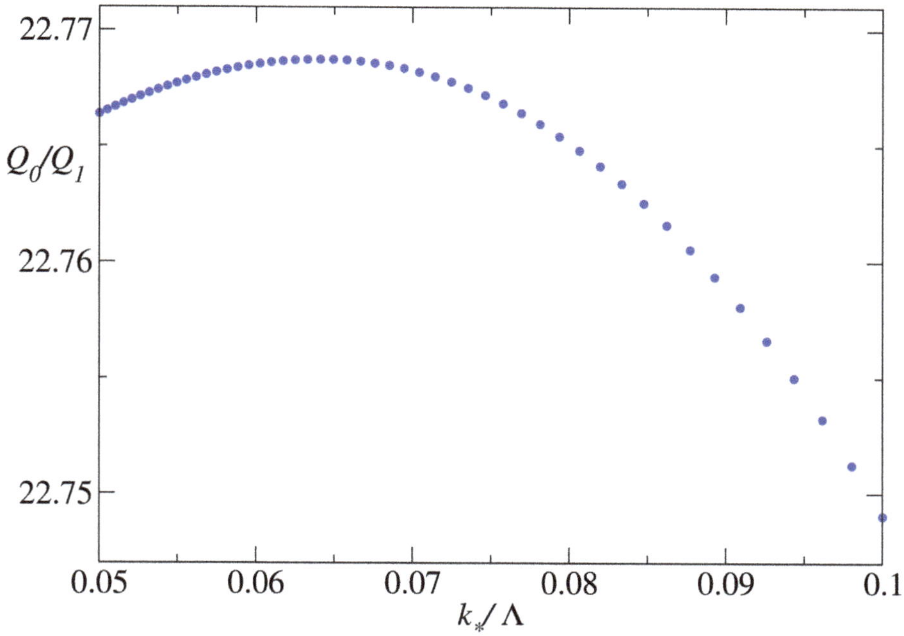

Figure 7.1. Numerical calculation of the ratio between ground state binding momentum and that of the first excited state using Danilov–Pricoupenko filtering, as a function of (inverse) cutoff.

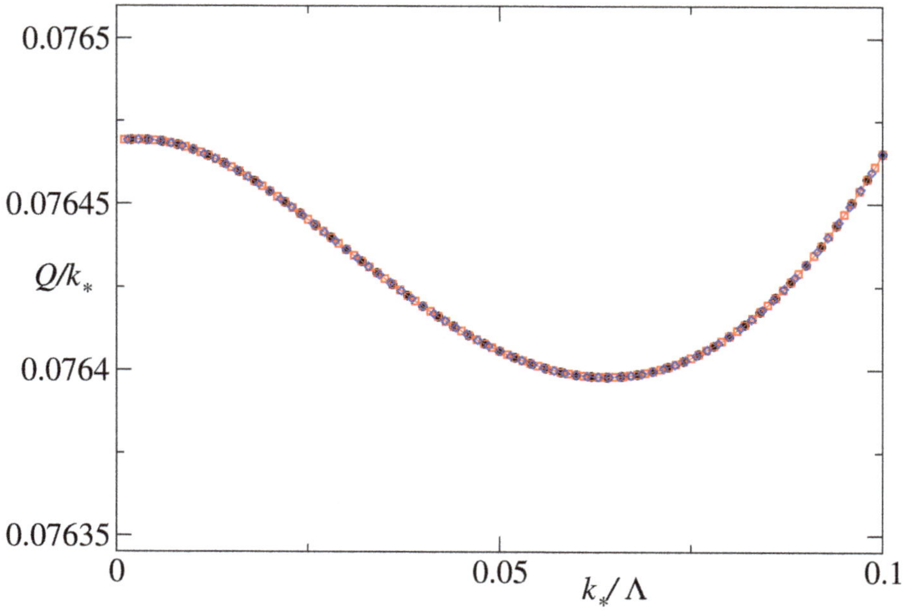

Figure 7.2. Numerical calculation of the ground state binding momentum in units of the three-body parameter k_*, using Danilov–Pricoupenko filtering, as a function of (inverse) cutoff.

Hence, the Skornyakov–Ter-Martirosyan equation including a contact three-body interaction in three spatial dimensions ($D = 3$) is given by

$$M(k) = 2\mathcal{T}\left(-\frac{\hbar^2}{m}(Q^2 + 3k^2/4)\right) \times \\ \frac{m}{\hbar^2 k}\int_0^\infty \frac{dq}{(2\pi)^2}\left[\mathcal{K}(k, q) + 3\frac{\hbar^2 qk}{m}T_U(E)I(E, k)I(E, q)\right]qM(q). \tag{7.80}$$

Above, T_U is the three-body T-matrix in the absence of two-particle interactions (see equation (7.7)), and the kernel $\mathcal{K}$ is given below equation (7.41).

If we leave the three-body interaction strength, typically denoted D_0, in bare form, then with a hard cutoff Λ we have $T_U \propto \Lambda^{-4}$, while $I(E, k) \propto \Lambda$. Let us look at this in some more detail in the following. The three-body T-matrix T_U is easily calculated from its Lippmann–Schwinger equation as

$$\frac{1}{T_U(z)} = \frac{1}{D_0} - \int \frac{d\mathbf{q}}{(2\pi)^3}\int \frac{d\mathbf{q}'}{(2\pi)^3}\frac{1}{z - \epsilon(\mathbf{q}) - \epsilon(\mathbf{q}') - \epsilon(\mathbf{q} + \mathbf{q}')}. \tag{7.81}$$

The integral in equation (7.81) is given by $\alpha\Lambda^4 + O(\Lambda^3)$, with α *energy-independent*. Meanwhile, $I(E, k) = \beta\Lambda$, with β energy-independent, too. Defining $3\hbar^2\alpha^{-1}\beta^2/m \equiv \gamma$, we obtain, to leading order

$$M(k) = 2\mathcal{T}\left(-\frac{\hbar^2}{m}\left(Q^2 + 3k^2/4\right)\right)\frac{m}{\hbar^2 k}\int_0^\Lambda \frac{dq}{(2\pi)^2}\left[\mathcal{K}\left(k, q\right) + qk\frac{\gamma}{\Lambda^2}\right]qM(q). \tag{7.82}$$

This is exactly what was obtained in reference [8], derived here in just a few lines for the joy of our diagram-averse readership. Redefining, as we did earlier, $f(k) = kM(k)$, the above equation reduces to

$$f(k) = 2\mathcal{T}\left(-\frac{\hbar^2}{m}\left(Q^2 + 3k^2/4\right)\right)\frac{m}{\hbar^2}\int_0^\Lambda \frac{dq}{(2\pi)^2}\left[\mathcal{K}\left(k, q\right) + qk\frac{\gamma}{\Lambda^2}\right]f(q). \tag{7.83}$$

We now investigate the integral equation (7.83). Its analysis, as opposed to its simple derivation, is very interesting and yields some very important insights into the low-energy theory of Efimov states. The goal of γ, which is made cutoff-dependent and tuneable (more on *why* this can be done later), will be to make the low-energy behaviour of $f(k)$ independent of the high-energy degrees of freedom. That is, we want the low-momentum limit of $f(k)$ to be independent of the sharp momentum cutoff Λ. We know already, from equation (7.54), that $f(k)$ should have a form similar to

$$f(k) \propto \frac{1}{k}\sin\left(s_0 \operatorname{arcsinh}\left(\frac{1}{2}\frac{k}{\Lambda_*}\right)\right), \tag{7.84}$$

with Λ_* some parameter with dimensions of momentum to make the argument of the hyperbolic function dimensionless. Hence, we aim at making the following quantity $A(\Lambda)$ independent of Λ,

$$A(\Lambda) = \int_0^\Lambda dq \left[\ln \left| \frac{Q^2 + k^2 + q^2 - kq}{Q^2 + k^2 + q^2 + kq} \right| + qk\frac{\gamma(\Lambda)}{\Lambda^2} \right] f(q), \tag{7.85}$$

where $f(q)$ is given by equation (7.84). A significant simplification of the analysis comes from the large-q expansion of the logarithm in equation (7.85) which, for $q \gg k$, is approximated as

$$\ln \left| \frac{Q^2 + k^2 + q^2 - kq}{Q^2 + k^2 + q^2 + kq} \right| \approx -\frac{2k}{q}, \quad q \gg k. \tag{7.86}$$

The next simplification at large values of q is obtained by replacing the arcsinh in equation (7.84) with the logarithm of twice the argument, that is,

$$f(q) \approx \frac{1}{q} \sin(s_0 \ln(q/\Lambda_*)), \quad q \gg \Lambda_*. \tag{7.87}$$

Using the two approximations (7.86) and (7.87), the leading term (in Λ), which we call $\delta A(\Lambda)$, of $A(\Lambda)$, equation (7.85), is obtained as

$$\delta A(\Lambda) \approx k \int^\Lambda dq \left[\frac{-2}{q^2} + \frac{\gamma(\Lambda)}{\Lambda^2} \right] \sin\left(s_0 \ln\left(\frac{q}{\Lambda_*} \right)\right). \tag{7.88}$$

To ensure cutoff independence, we choose yet another value of the cutoff, Λ', satisfying $\Lambda' > \Lambda$, and calculate $\gamma(\Lambda)$ so that

$$B(\Lambda', \Lambda) \equiv \frac{\delta A(\Lambda') - \delta A(\Lambda)}{k} = 0. \tag{7.89}$$

Inserting equation (7.88) into equation (7.89) above, we obtain

$$\begin{aligned} B(\Lambda, \Lambda') = & \int_\Lambda^{\Lambda'} dq \left[\frac{-2}{q^2} + \frac{\gamma(\Lambda')}{\Lambda'^2} \right] \sin\left(s_0 \ln\left(\frac{q}{\Lambda_*} \right)\right) \\ & + \int^\Lambda dq \left[\frac{\gamma(\Lambda')}{\Lambda'^2} - \frac{\gamma(\Lambda)}{\Lambda^2} \right] \sin\left(s_0 \ln\left(\frac{q}{\Lambda_*} \right)\right). \end{aligned} \tag{7.90}$$

We basically need to calculate two rather straightforward integrals, the first of which is

$$\begin{aligned} I_1(\Lambda) &= -2 \int^\Lambda dq \frac{1}{q^2} \sin\left(s_0 \ln\left(\frac{q}{\Lambda_*} \right)\right) \\ &= 2\frac{1}{\Lambda}\frac{1}{1 + s_0^2}[\sin(s_0 \ln(\Lambda/\Lambda_*)) + s_0 \cos(s_0 \ln(\Lambda/\Lambda_*))], \end{aligned} \tag{7.91}$$

while the second one is simply

$$
\begin{aligned}
I_2(\Lambda) &= \int^{\Lambda} dq \sin\left(s_0 \ln\left(\frac{q}{\Lambda_*}\right)\right) \\
&= \Lambda \frac{1}{1+s_0^2}[\sin(s_0 \ln(\Lambda/\Lambda_*)) - s_0 \cos(s_0 \ln(\Lambda/\Lambda_*))].
\end{aligned} \tag{7.92}
$$

We may simplify the results of the two integrals by rewriting $\sin x \pm s_0 \cos x$ as $\sin(x+\delta)/\cos\delta$, with $\tan\delta = s_0$, and using that $1 + s_0^2 = \sec^2\delta$. We then easily see that

$$
I_1(\Lambda) = 2\frac{1}{\Lambda}\frac{1}{\sqrt{1+s_0^2}} \sin(s_0 \ln(\Lambda/\Lambda_*) + \arctan(s_0)), \tag{7.93}
$$

$$
I_2(\Lambda) = \Lambda \frac{1}{\sqrt{1+s_0^2}} \sin(s_0 \ln(\Lambda/\Lambda_*) - \arctan(s_0)). \tag{7.94}
$$

Inserting equations (7.93) and (7.94) into equation (7.90), and setting $B(\Lambda, \Lambda') = 0$, gives us the cutoff dependency of $\gamma(\Lambda)$,

$$
\gamma(\Lambda) = -2\frac{\sin(s_0 \ln(\Lambda/\Lambda_*) + \arctan(s_0))}{\sin(s_0 \ln(\Lambda/\Lambda_*) - \arctan(s_0))}. \tag{7.95}
$$

We see that, indeed, γ is of $O(1)$, but has a non-trivial, periodic dependence on the UV cutoff Λ. Also note that with the above procedure we have neglected contributions of order smaller than Λ^{-1}, that is, we have removed the leading-order cutoff dependence.

We also need to see whether γ can be tuned at all, a problem that is not tackled nor rigorously resolved in reference [8]. Recall that γ came originally from leaving the three-body bare coupling constant D_0 fixed to a finite value, see equation (7.83) and the subsequent discussion. Then, as $\Lambda \gg \Lambda_*$, γ becomes essentially independent of D_0: the leading-order correction, which does depends on D_0 if this is finite, is of $O(\Lambda^{-4})$. Fortunately, in contrast to reference [8], our approach here is so simple that it is quite standard and transparent to see how γ can be tuned at will. All we need to do is to set

$$
D_0(\Lambda) = \frac{d_R}{\Lambda^4}, \tag{7.96}
$$

and consequently γ becomes

$$
\gamma = \frac{3\hbar^2\beta^2}{m(d_R + \alpha)}. \tag{7.97}
$$

We conclude the discussion on this method by pointing out how this type of three-body interaction, being trivial on its own, reproduces highly non-trivial physics when put in combination with two-body interactions.

It is important to see how the different methods (usual renormalization, Danilov–Pricoupenko filtering, and addition of a three-body counterterm) perform when a large but finite momentum cutoff is placed. This is especially relevant since the method that uses an explicit three-body force has been obtained using a 'cutoff-heavy' derivation. We anticipate that the usual renormalization and Danilov–Pricoupenko filtering have essentially identical convergence properties with finite cutoffs, so we only plot numerical results using the latter method. Firstly, we should be aware of what cutoff independence should mean. Since we have worked in the unitary limit, plotting the ground state binding momentum as a function of the three-body parameter, in dimensionless form, implies that we plot Q/k_* (Q/Λ_*) as a function of k_*/Λ (Λ_*/Λ). This is because the only two momentum scales we can choose are k_* (Λ_*) and Λ itself. Therefore, the ground state binding momentum, if it should be (almost) independent of the cutoff, must have the form $Q = \beta k_*$ ($Q = \lambda\Lambda_*$) with β (λ) a dimensionless, positive number. In the case of Danilov–Pricoupenko filtering, we know that this number is, as $\Lambda \to \infty$, $\beta \approx 0.0765$, see equation (7.74). In figure 7.2, we see that convergence is very good as $k_*/\Lambda \to 0$, and even for moderate to large values of it, the result is in very good agreement with the exact result of equation (7.74). For the case of a three-body counterterm, the result of a numerical calculation is plotted in figure 7.3. Convergence is slightly worse (yet still impressive!) than in the case of Danilov–Pricoupenko filtering. However, this is partly due to the fact that our estimate of the three-body constant $\gamma(\Lambda)$, equation (7.95), is only approximate (it gets better as Λ grows, obviously). An accurate numerical calculation of $\gamma(\Lambda)$ yields improved results, see reference [8]. As for the excited states, it is easy to verify numerically that all methods give excellent scaling properties, namely that the binding momenta Q_n ($n = 0, 1, \ldots$) satisfy

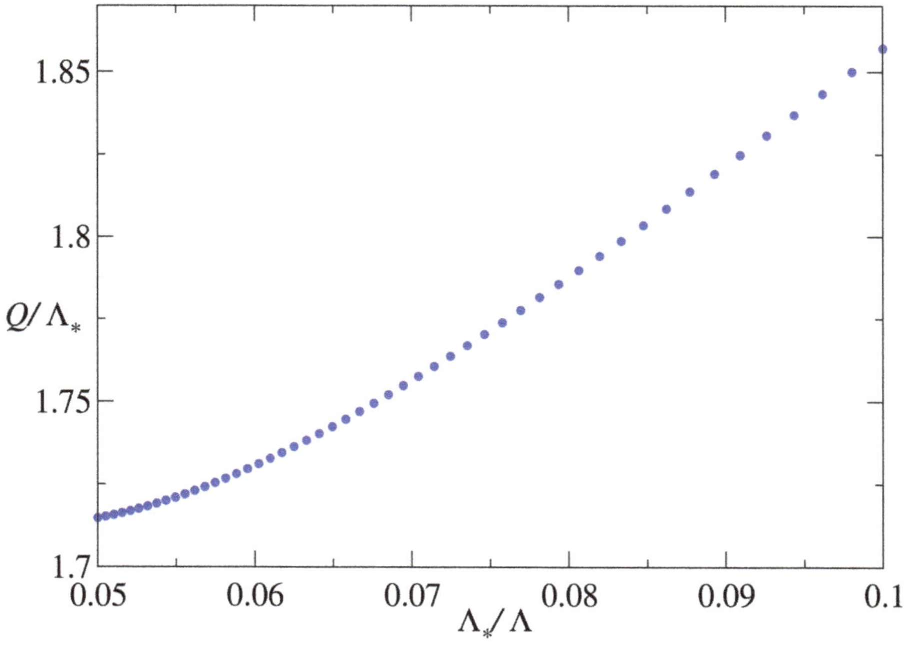

Figure 7.3. Numerical calculation of the ground state binding momentum in units of the three-body parameter Λ_*, using a three-body counterterm, as a function of (inverse) cutoff.

$$\frac{Q_n}{Q_{n+1}} = e^{\pi/s_0} \approx 22.7. \tag{7.98}$$

We plot this quantity for Danilov–Pricoupenko filtering in figure 7.1, where we indeed see that the scaling is quite accurately satisfied even with a finite cutoff.

7.1.3 Dimer-atom collisions

We have gone a long way with the three-boson problem, especially in three dimensions, regarding the bound states and the Efimov effect. That problem only requires solving the homogeneous version of the Faddeev equations. In contrast with the two-body problem, where the only two available channels are the scattering (positive energy) and bound state (negative energy) sectors, the three-body problem has extra channels. If the three particles are distinguishable, and labelled i, j and ℓ, then the channels and their main relevance are as follows:

- $i + j + \ell \to i + j + \ell$, or $3 \to 3$, that is, three free particles scattering elastically. This channel is most relevant for atomic quantum gas phases in which all atoms are 'free' and colliding. The energy in this channel is simply the sum of the kinetic energies of the particles.
- $(ij) + \ell \to (ij) + \ell$, that is, a bound state of particles i and j elastically colliding with particle ℓ, which is unbound. The energy of the channel is $E_{ij} + \hbar^2 K_{ij}^2/2M_{ij} + \hbar^2 k_\ell^2/2m_\ell$, with obvious notation. There are three channels of this type, corresponding to the three possible, different bound states of two particles.
- $(ij) + \ell \to (i\ell) + j$, that is, a rearrangement collision where the bound state of particles i and j turns into a bound state between particles i and ℓ. The energy of this channel is identical to the previous one. There are also three channels of this type.

  The two last types of channel will be denoted $2 + 1 \to 2 + 1$.
- $(ij) + \ell \to i + j + \ell$, that is, a breakup channel where a two-body bound state and a free particle collide and the bound state breaks up. For this to happen, the energy must be above the threshold. There are also three of these channels, which we shall call $2 + 1 \to 3$. Obviously if this channel is allowed, the inverse $3 \to 2 + 1$ is also allowed (and is called the recombination channel).
- (ijk), that is, the three-body bound state channel, which we have already explored.

The number of channels is drastically reduced if the particles are indistinguishable, since the $2 + 1 \to 2 + 1$ processes all collapse into a single channel, namely the elastic scattering of a bound state and a free particle. The same is true for the breakup channel, which is in this case unique. Therefore, for identical particles there are three collisional channels and one bound state channel.

In this section, we only introduce some of the formalism for $2 + 1 \to 2 + 1$ elastic scattering channel, with distinguishable particles. We follow Faddeev's original article [9], albeit with more modern notation. Without further ado, we shall write $|\Phi_n\rangle$ for the asymptotic basis functions of channel n. The final goal of any scattering calculation is the obtention of the S-matrix or, more concretely, the transition amplitudes from an initial state $|\phi_i\rangle$ to a final state $|\phi_f\rangle$. If the full scattering state is called $|\psi\rangle$, then the transition amplitude is given by

$$T_{fi} = \langle \phi_f | V^f \psi \rangle. \tag{7.99}$$

Above, V^f is the part of the interaction not taken into account in the state $|\phi_f\rangle$. For instance, if the transition is from some state $|\phi_i\rangle$ to a free three-body state, then V^f is the entire interaction, while if the transition we are interested in is towards a state comprised of a dimer (ij) and a single-particle ℓ, then V^f is given by

$$V^f = V - V^\ell = V^i + V^j = V_{j\ell} + V_{i\ell}, \tag{7.100}$$

where the last equality is only true if there are only pairwise interactions. Coming back to the asymptotic basis function $|\Phi_n\rangle$, the scattering state can be obtained as

$$|\psi_n\rangle = i\eta G(E_n + i\eta)|\Phi_n\rangle, \tag{7.101}$$

where the limit $\eta \to 0^+$ is implicitly assumed, as usual, and E_n is the energy of $|\psi_n\rangle$. Let us prove equation (7.101). The Green's operator satisfies equation (1.64). Particularizing to $z = E_n + i\eta$ and applying both sides of the equation on $|\Phi_n\rangle$, and multiplying these by $i\eta$, we have

$$i\eta G(E_n + i\eta)|\Phi_n\rangle = i\eta G_0(E_n + i\eta)|\Phi_n\rangle + i\eta G_0(E_n + i\eta) V G(E_n + i\eta)|\Phi_n\rangle. \tag{7.102}$$

Since $H_0|\Phi_n\rangle = E_n|\Phi_n\rangle$, by definition, then the first term on the right-hand-side of equation (7.102) is just $|\Phi_n\rangle$. The function $|\tilde{\psi}\rangle \equiv i\eta G(E_n + i\eta)|\Phi_n\rangle$, therefore, satisfies the equation

$$|\tilde{\psi}\rangle = |\Phi_n\rangle + G_0(E_n + i\eta) V |\tilde{\psi}\rangle, \tag{7.103}$$

which is the equation satisfied by a scattering state. Hence, we conclude that $|\tilde{\psi}\rangle = |\psi_n\rangle$, and equation (7.101) is proved. The solution of a 2 + 1 elastic collision, from here, is not much more involved than the three-body bound state problem, and we leave the details to the interested reader (see references [9, 10]).

7.1.4 Induced three-body interactions in one dimension

Three-body interactions—those that cannot be separated as sums of potentials depending only on the coordinates and quantum numbers of pairs of particles—are typically not discussed in most textbooks on quantum mechanics or atomic and molecular physics, nor in research articles in these fields for that matter. There is no fundamental reason why these should not play a role, and in fact they are quite necessary in nuclear physics. The question of whether three-body forces are genuine or effective was also the subject of intense investigation in the 20th century, especially in the context of nuclear physics. Any discussion on this point leads nowhere, unless the constituent particles are actually fundamental—neither protons, neutrons nor, of course, atomic nuclei, not to mention atoms and molecules, are fundamental. Therefore, even if only genuine two-body interactions between fundamental constituents were present, effective three-body interactions would arise, or would be indistinguishable from genuine ones. From a technical perspective, this issue is related with the off-shell elements of the two-body transition operator. For three particles, upon iteration of the Faddeev equations, these off-shell elements generate connected three-body (on-shell) amplitudes that might as well be reproduced using on-shell three-body forces rather than off-shell two-body T-matrices [11].

In this section, we study the simplest scenario, in quantum mechanics, where three-body interactions may emerge from pairwise interactions, and be essentially isolated from those at low energies. In particular, we consider three spinless bosons in one spatial dimension ($D = 1$), interacting via short-range, pairwise potentials of arbitrary shape and strength. To be able to isolate emergent, or effective, three-body interactions from two-body effects, we set the two-body subsystems on resonance (infinite scattering length, $1/a \to 0$). In one dimension, infinite scattering length turns two bosons into effectively non-interacting at low energies. For simplicity, we consider $1/a \to 0^-$, which allows us to choose a two-body potential that supports no bound states. As such, the ground state can be a $3 \to 3$ scattering state.

Before digging into the problem, we need to discuss low-energy three-body interactions in one dimension. The simplest type of three-body potential is 'spherically' symmetric. For a local interaction V_3, this means that in the position representation it takes the form

$$V_3(x_1, x_2, x_3) = F(\rho), \tag{7.104}$$

that is, it is a function of the three-body radius ρ only. In one dimension, the radius is given by

$$\rho = \sqrt{x^2 + y^2}, \tag{7.105}$$

where x and y are Jacobi coordinates, of the form

$$x = \frac{x_1 - x_2}{\sqrt{2}}, \tag{7.106}$$

$$y = \sqrt{\frac{2}{3}}\left[x_3 - \frac{x_1 + x_2}{2}\right]. \tag{7.107}$$

Our problem is much more conveniently handled in the momentum representation. One such local, spherically-symmetric potential takes the form

$$\langle \vec{k}'|V_3|\vec{k}\rangle = 2\pi\delta(K - K')f(q_\rho^2) \equiv 2\pi\delta(K - K')v(\vec{k}', \vec{k}), \tag{7.108}$$

where $q_\rho = (q_x^2 + q_y^2)^{1/2}$, with q_x and q_y given by changing $x \to q_x$, $y \to q_y$, and $x_i \to k_i - k_i'$ ($i = 1, 2, 3$) in equations (7.106) and (7.107). Low-energy interactions, or three-body EFT, in one dimension, is naturally derived by expanding f, equation

(7.108), in powers of q_ρ^2. To leading order, the three-boson interaction is a constant, namely

$$v_{\mathrm{LO}}(\vec{k}', \vec{k}) = g_3, \tag{7.109}$$

where we have obviated the momentum conservation for simplicity. For indistinguishable fermions, the leading-order interaction shows up at $O(q_\rho^6)$, and is given by

$$v_{\mathrm{LO}}^{\mathrm{F}}(\vec{k}', \vec{k}) = g_3^{\mathrm{F}} k'_{ij} k'_{i\ell} k'_{j\ell} k_{ij} k_{i\ell} k_{j\ell}. \tag{7.110}$$

The three-boson interaction in equation (7.109) is so simple that it leads to an exactly solvable problem for three particles. To see this, we solve the Lippmann–Schwinger equation for the three-body transition operator $T(z)$. Note that, in this case, the Lippmann–Schwinger equation does not require Faddeev decomposition since the interaction (7.108) is connected. After separation of the centre-of-mass momentum, the T-matrix becomes a constant in the momentum representation. The solution is standard, and we obtain for $K = 0$

$$T(z) = \frac{1}{g_3^{-1} - I(z)}, \tag{7.111}$$

with

$$I(z) = \int \frac{dq_1}{2\pi} \int \frac{dq_2}{2\pi} \frac{1}{z - \frac{\hbar^2}{m}\left[q_1^2 + q_2^2 + q_1 q_2\right]}. \tag{7.112}$$

The simplest way to calculate $I(z)$ above is to use Jacobi coordinates, followed by a transformation to polar coordinates. From the definition of the Jacobi coordinates, we obtain $q_1 = q_1(q_x, q_y)$ and $q_2(q_x, q_y)$ as

$$q_1 = \frac{1}{\sqrt{2}} q_x - \frac{1}{\sqrt{6}} q_y, \tag{7.113}$$

$$q_2 = -\frac{1}{\sqrt{2}} q_x - \frac{1}{\sqrt{6}} q_y. \tag{7.114}$$

The Jacobian J of the transformation takes the value $J = 1/\sqrt{3}$, and the integral $I(z)$ in equation (7.112) changes to

$$I(z) = \frac{1}{\sqrt{3}} \int \frac{dq_x}{2\pi} \frac{dq_y}{2\pi} \frac{1}{z - \frac{\hbar^2}{2m}(q_x^2 + q_y^2)} = \frac{1}{\sqrt{3}} \int_0^\Lambda \frac{dq_\rho}{2\pi} \frac{q_\rho}{z - \frac{\hbar^2}{2m} q_\rho^2}, \tag{7.115}$$

where we have introduced a hard cutoff for q_ρ, since the integral is UV-divergent. For scattering states, we set $z = \hbar^2 k^2 / 2m + i\eta$ ($\eta \to 0^+$), and easily calculate $I(z)$, which takes the value

$$I(\hbar^2k^2/2m + i\eta) = -\frac{m}{\sqrt{3}\pi\hbar^2}\left[\ln\left(\frac{\Lambda}{k}\right) + i\frac{\pi}{2}\right]. \tag{7.116}$$

We insert equation (7.116) into equation (7.111), and renormalize the T-matrix such that it is purely imaginary for $k = Q_*$, with Q_* the fixed renormalization scale. The bare coupling constant then takes the simple form

$$\frac{1}{g_3} = \frac{m}{\sqrt{3}\pi\hbar^2}\ln\left|\frac{Q_*}{\Lambda}\right|, \tag{7.117}$$

and the final form of the renormalized T-matrix becomes

$$T(\hbar^2k^2/2m + i\eta) = \frac{\sqrt{3}\pi\hbar^2}{m}\left[\ln\left|\frac{Q_*}{k}\right| + i\frac{\pi}{2}\right]^{-1}. \tag{7.118}$$

The renormalization scale Q_* can be interpreted in two different ways. Firstly, if the (original) three-body interaction is attractive, then $\hbar^2Q_*^2/2m$ is the binding energy of the allowed three-body bound state. This is seen by analytically continuing equation (7.118) to imaginary values of $k \equiv i\lambda$ ($\lambda > 0$). The principal value of the logarithm in equation (7.118) is given by $\ln|Q_*/k| = \ln|Q_*/\lambda| - i\pi/2$ and, therefore, T is singular —and the singularity is a simple pole—for $z = -\hbar^2Q_*^2/2m$. The second interpretation needs to be taken for repulsive interactions: since the pole in the T-matrix is always present, for repulsion the pole is spurious and must not be considered. This phenomenon we shall call Landau pole from here on.

To show that three-body interactions emerge, we shall focus on the repulsive regime. While the Faddeev equations can be used to show that such interaction emerges when $1/a \to 0$, this is rather complicated. Therefore, we begin by searching for signatures of the three-body interaction following a different route. What we will do is numerically solve the three-boson problem with some semi-realistic two-body interaction, and $1/a \to 0^-$ with no bound states, in a finite box of length L with periodic boundary conditions. If the emergent three-body interaction is repulsive, then the ground state energy E_0 of the finite system will be shifted upwards, that is, $E_0 > 0$. The effective three-body interaction parameters can be inferred from the energy shifts as functions of L, just as we did for two particles in chapter 2. We begin by relating Q_* with the ground state energy. Our starting point is equation (2.115), which is still valid for three particles, but all operators refer to three-body operators instead. We write it again as

$$\hat{r}(E) = \hat{T}(z) + \hat{T}(z)\delta\tilde{G}_0(L, E, z)\hat{r}(E). \tag{7.119}$$

Just as we did in chapter 2, we use the R-matrix instead of the T-matrix, which is more convenient as it is purely real. The R-matrix is obtained from the T-matrix by taking the real part of the inverse of the T-matrix in equation (7.118), that is, the R-matrix is given by

$$R(\hbar^2 k^2/2m) = \frac{\sqrt{3}\pi\hbar^2}{m} \frac{1}{\ln\left|\dfrac{Q_*}{k}\right|}. \tag{7.120}$$

In equation (7.119), using the R-matrix allows us to make the replacement $z = E + i\eta \to E$. We shall also use the notation of chapter 2 for the plane-wave basis, which uses uppercase (lowercase) symbols to denote infinite (finite) volume so, for instance,

$$\langle x|q\rangle = \frac{1}{L^{1/2}} e^{iqx}, \tag{7.121}$$

$$\langle x|Q\rangle = e^{iQx}. \tag{7.122}$$

We also have $\langle q'|q\rangle = \delta_{q',q}$ and $\langle Q'|Q\rangle = (2\pi)\delta(Q - Q')$ (see chapter 2 for more details). We will implicitly assume that the total momentum vanishes. With the three-body notation for the basis functions, equation (7.23), and eliminating the centre-of-mass momentum, we have the following relations, given an operator $\hat{O}$,

$$\langle \vec{k}'|\hat{O}|\vec{k}\rangle = \frac{1}{L^2}\langle \vec{Q}'|\hat{O}|\vec{Q}\rangle, \tag{7.123}$$

$$\langle \vec{k}|\hat{O}|\vec{Q}\rangle = \frac{1}{L}\langle \vec{K}|\hat{O}|\vec{Q}\rangle. \tag{7.124}$$

To avoid any confusion, note that when we work in the infinite size limit we always use un-normalized plane waves, and that is the basis in which the T-matrices and momentum-represented interaction potentials are always written. Without further ado, we use all the above notational rules, and the fact that the R-matrix, equation (7.120), is a constant (independent of momentum), in equation (7.119), obtaining

$$\begin{aligned}\langle \vec{k}'|\hat{r}(E)|\vec{k}\rangle = \frac{R(E)}{L^2} &+ \frac{R(E)}{L^2}\sum_{\mathbf{q}\neq\mathbf{k}} \frac{\langle \vec{q}|\hat{r}(E)|\vec{k}\rangle}{E - \epsilon(q_x) - \epsilon(q_y) - \epsilon(q_x + q_y)} \\ &- \frac{R(E)}{L}\mathcal{P}\int \frac{d^2Q}{(2\pi)^2} \frac{\langle \vec{Q}|\hat{r}(E)|\vec{k}\rangle}{E - \epsilon(Q_x) - \epsilon(Q_y) - \epsilon(Q_x + Q_y)}.\end{aligned} \tag{7.125}$$

From the above equation, it clearly holds that $\langle \vec{k}'|\hat{r}(E)|\vec{k}\rangle$ is a constant, and we call it simply $r(E)$. Further using equations (7.123) and (7.124), setting $\mathbf{k} = 0$, and solving for $r(E)$, we obtain its explicit form as

$$r(E) = \frac{L^{-2}}{\dfrac{1}{R(E)} - \left[\dfrac{1}{L^2}\sum_{\vec{q}\neq 0} \dfrac{1}{E - \epsilon(q_x) - \epsilon(q_y) - \epsilon(q_x + q_y)} - \mathcal{P}\int \dfrac{d^2Q}{(2\pi)^2} \dfrac{1}{E - \epsilon(Q_x) - \epsilon(Q_y) - \epsilon(Q_x + Q_y)}\right]}. \tag{7.126}$$

We also know, from the Bethe–Goldstone equation, equation (2.9), that $r(E) = E$. Writing $E = \hbar^2\lambda^2/2mL^2$, we obtain from equation (7.125) and the explicit expression for the R-matrix, equation (7.120), the following equation for λ,

$$\frac{1}{\lambda^2} = \frac{\ln\left|\frac{(Q_*L)^2}{\lambda^2}\right|}{2\sqrt{3}\pi} - 2\left[\frac{1}{L^2}\sum_{\vec{q}\neq 0}\frac{1}{\frac{\lambda^2}{L^2} - 2(q_x^2 + q_y^{2*}q_xq_y)} - \mathcal{P}\int\frac{d^2Q}{(2\pi)^2}\frac{1}{\frac{\lambda^2}{L^2} - 2(Q_x^2 + Q_y^{2*}Q_xQ_y)}\right]. \tag{7.127}$$

The above equation determines the energy, from λ, fully non-perturbatively. The situation regarding the weak-coupling limit is similar to the two-dimensional case with two-body LO EFT interactions, where the coupling constant, being logarithmic, can be essentially chosen arbitrarily [12]. The best one can do somewhat analytically, at extremely small coupling constants, or extremely long lengths L, is to neglect the second line in equation (7.127), and obtain an approximate value for λ, given by the solution to

$$\frac{1}{\lambda} \approx \frac{\ln\left|\frac{(Q_*L)^2}{\lambda^2}\right|}{2\sqrt{3}\pi}, \quad Q_*L \to \infty. \tag{7.128}$$

The problem here is that the term in the second line of equation (7.127) is of $O(1)$, which is the same order as $1/\lambda^2$ (see reference [13]). Hence, for the weak-coupling limit to be truly achieved, we need $(Q_*L) = O(\exp(\kappa L))$, with κ a momentum scale of $O(1)$. This requires unnaturally large, size-dependent Q_*, $Q_* \sim \exp(\kappa L)/L$. Therefore, the best option is to solve equation (7.128) non-perturbatively, which is numerically not very demanding. In figure 7.4, we show the value of λ as we vary Q_*L. Due to the logarithmic dependence of the T-matrix on Q_*L, the energy changes very slowly as we vary Q_* or, for that matter, the length of the system. This is exactly what happens on resonance (from the left), with $1/a \to 0^-$, for a three-body system interacting via pairwise potentials (see reference [13]).

7.1.5 Three-boson problem on the lattice: Mattis equation

Up until this point in this chapter, we have considered non-relativistic, Galilean bosons. From a practical standpoint, this simply means that we have been able to leave the conserved total momentum $\mathbf{K} = \mathbf{k}_i + \mathbf{k}_j + \mathbf{k}_\ell$ fixed to an arbitrary value that we have set to be zero, for simplicity. All the physics was invariant with respect to a Galilean boost. In this section, however, we study the three-body problem, for identical bosons, on a tight-binding lattice. As we saw in the case of two particles in the previous chapter, a tight-binding model does not exhibit Galilean invariance, since the single-particle spectrum is necessarily not quadratic in the (quasi-) momentum. This has interesting effects, as we have seen, and pairs of particles

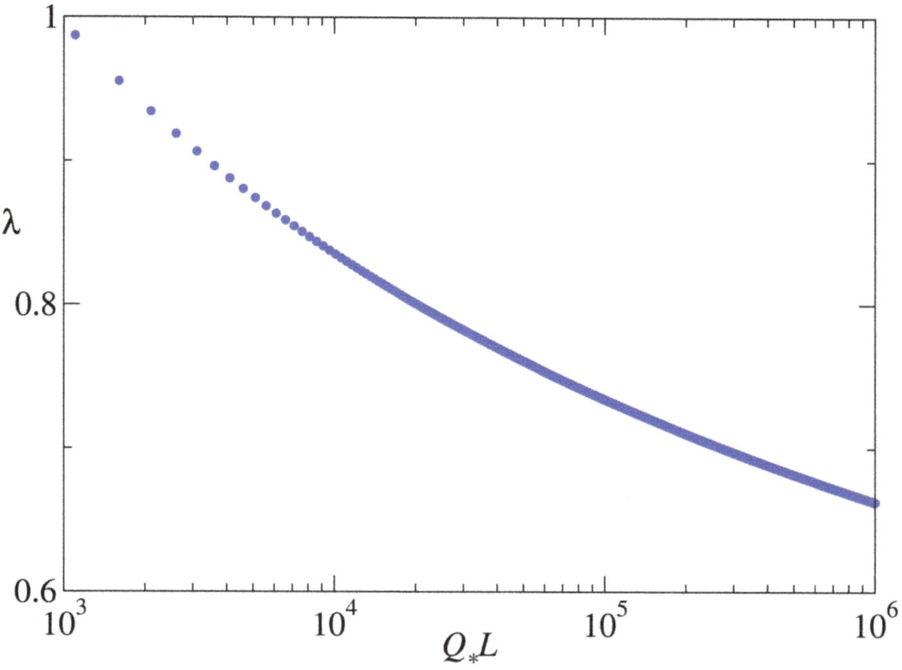

Figure 7.4. Dimensionless momentum shift from a one-dimensional three-body interaction in one dimension, as a function of the dimensionless momentum scale Q_*L. Note that the abscissa is in logarithmic scale.

with different total quasi-momenta behave differently. A particular phenomenon associated with the lack of Galilean invariance is the so-called Mattis–Gallinar effect: pairs of particles with total momenta near the edge of the Brillouin zone bind tighter than otherwise.

For this section, we shall partially follow reference [14], although the reader is also encouraged to carefully read the original source [15]. The derivation of the three-body equations does not differ in any way, in operator form, from what we followed for the Faddeev equations. In fact, the resulting scattering equations are identical, and given by equations (7.15), (7.16). The same is true for the bound states. Here, however, the Hamiltonian is different. We shall take a hypercubic lattice in D-dimensions with minimal (nearest-neighbour) tunnelling, with rate J (>0), and leading-order two-body interactions with strength U. The $U \rightarrow -U$ symmetry is present, which allows us to particularize, for simplicity, to attractive interactions ($U < 0$). The lattice spacing is called d. The single-particle dispersion takes the following form, well known by now,

$$\epsilon(\mathbf{k}) = -2J\sum_{\alpha=1}^{D}(\cos(k_\alpha d) - 1). \tag{7.129}$$

Above, $\alpha = 1$ corresponds to x, $\alpha = 2$ to y, and so on. The derivation of the bound state integral equation is no different from that in the continuum, except that we cannot arbitrarily set the total momentum to zero, since our system is no longer

Galilean relativistic. The equation for the M function, which we write $M_{\mathbf{K}}$, due to its parametric dependence on the total momentum $\mathbf{K}$, reads

$$M_{\mathbf{K}}(z; \mathbf{k}) = 2\mathcal{T}_{\mathbf{K}}(z, \mathbf{k}) \int_{\mathrm{BZ}} \frac{d\mathbf{q}}{\Omega} \frac{M_{\mathbf{K}}(z, \mathbf{q})}{z - \epsilon(\mathbf{k}) - \epsilon(\mathbf{q}) - \epsilon(\mathbf{K} - \mathbf{k} - \mathbf{q})}. \tag{7.130}$$

Above, $\Omega = (2\pi/d)^D$ is the volume (in reciprocal space) of the Brillouin zone, and the integral is over the Brillouin zone $(-\pi/d, \pi/d]^D$. For contact interactions, the off-shell T-matrix $\mathcal{T}_{\mathbf{K}}(\mathbf{k})$ is a solution to

$$\mathcal{T}_{\mathbf{K}}(z, \mathbf{k}) = U + U \int_{\mathrm{BZ}} \frac{d\mathbf{q}}{(2\pi)^D} \frac{\mathcal{T}_{\mathbf{K}}(z, \mathbf{k})}{z - \epsilon(\mathbf{q}, \mathbf{k}, \mathbf{K} - \mathbf{k} - \mathbf{q})}. \tag{7.131}$$

Solving equation (7.130) numerically, at least in one spatial dimension, is straightforward. However, we can use a solution method for the ground state that is not available in the continuum, namely the strong-coupling expansion. Since the three-body dispersion relation is bounded from above, if the interaction is large enough, we can expand the Green's function in equation (7.130) as

$$\frac{1}{z - \epsilon} = \frac{1}{z} \frac{1}{1 - \epsilon/z} = \frac{1}{z}\left[1 + \frac{\epsilon}{z} + \frac{\epsilon^2}{z^2} + \cdots\right], \tag{7.132}$$

where ϵ is being used as shorthand for $\epsilon(\mathbf{k}) + \epsilon(\mathbf{q}) + \epsilon(\mathbf{K} - \mathbf{k} - \mathbf{q})$. The T-matrix, for large values of $|U|/J$, can also be expanded, with the leading-order term being $\mathcal{T}_{\mathbf{K}} = 2U/(1 - U/z)$. Therefore, the lowest-order approximation in the strong-coupling limit is given by the solution to the equation

$$M_{\mathbf{K}}(z; \mathbf{k}) = \frac{2U}{1 - U/z} \int_{\mathrm{BZ}} \frac{d\mathbf{q}}{\Omega} \frac{M_{\mathbf{K}}(z, \mathbf{q})}{z}. \tag{7.133}$$

Obviously, equation (7.133) is solved by a constant $M_{\mathbf{K}}$, and we have

$$z = 3U. \tag{7.134}$$

This result is quite simple to understand if we work in the position representation instead. Placing three bosons on the same site, with a large interaction $|U|/J$, gives to zeroth order $E = UN(N - 1)/2 = U3(3 - 1)/2 = 3U$. In problem 7, the reader is asked to continue the expansion to the following order.

7.2 Efimov effect for non-identical particles

In this section we consider mixtures of different types of particles. In particular, we study three particles of two different kinds, one being called heavy, with mass M and the other one light, with mass m. The reason for investigating this system, besides its actual physical importance, is that if two heavy particles with masses $M \gg m$ interact resonantly with the light particle, an approximate treatment, within the Born–Oppenheimer approach, yields a very transparent view on the Efimov effect. The resulting solution is exact (in this approximation), and we obtain the very same scale invariance, and breaking thereof down to discrete scale invariance, that we saw for three identical bosons. Here, we will follow reference [16].

7.2.1 Born–Oppenheimer approximation

We begin by considering the two heavy particles, with mass M, separated by a distance R, at fixed positions. These can be set to $\pm\mathbf{R}/2$ without loss of generality. A third, lighter particle with mass m, interacts with the two heavy particles via a (near-) resonant, short-range interaction, which at low energies is modelled by the LO EFT in the s-wave channel, characterized by the scattering length a, with $|a| \gg R$. The two heavy particles may be regarded as non-interacting with each other. This is the first step of the Born–Oppenheimer approximation, where one obtains the light-heavy-heavy bound state energy, $\mathcal{E}(R)$, as a function of the fixed separation between the two heavy particles. The second step consists of solving the two-body Schrödinger equation for the two heavy particles subjected to the Born–Oppenheimer potential, which is given by $\mathcal{E}(R)$.

We denote $\psi_{\mathbf{R}}(\mathbf{r})$ the eigenstates of the following Hamiltonian, for the itinerant light particle,

$$H_m = -\frac{\hbar^2}{2m}\nabla_{\mathbf{r}}^2 + V_{\mathrm{LO}}(\mathbf{r} - \mathbf{R}/2) + V_{\mathrm{LO}}(\mathbf{r} + \mathbf{R}/2), \tag{7.135}$$

where V_{LO} is the LO EFT s-wave interaction, which may be written as the Lee–Huang–Yang pseudopotential, namely

$$V_{\mathrm{LO}}(\mathbf{r}) = \frac{4\pi\hbar^2 a}{m}\delta(\mathbf{r})\frac{\partial}{\partial r}(r\ \cdot). \tag{7.136}$$

The above interaction implies the so-called Bethe–Peierls boundary condition on an eigenstate ψ, which reads

$$\psi(\mathbf{r} \to 0) \propto \frac{1}{a} - \frac{1}{r}. \tag{7.137}$$

The proportionality sign above is necessary since the overall normalization constant for ψ is *a priori* undetermined. The stationary Schrödinger equation for Hamiltonian H_m in the bound state sector has solutions of the general form

$$\psi_{\mathbf{R}}(\mathbf{r}) = A\frac{e^{-\lambda(R)|\mathbf{r}-\mathbf{R}/2|}}{|\mathbf{r} - \mathbf{R}/2|} + B\frac{e^{-\lambda(R)|\mathbf{r}+\mathbf{R}/2|}}{|\mathbf{r} + \mathbf{R}/2|}, \tag{7.138}$$

with A and B two constants whose relation B/A must be determined, while the bound state energy is given by

$$\mathcal{E}(R) = -\frac{\hbar^2[\lambda(R)]^2}{2m}. \tag{7.139}$$

Applying the Bethe–Peierls boundary conditions, equation (7.137), adapted to Hamiltonian (7.135), gives us, as $\mathbf{r} \to \pm\mathbf{R}/2$, the following two relations

$$\frac{1}{a} - \frac{1}{|\mathbf{r} - \mathbf{R}/2|} \propto \frac{A}{|\mathbf{r} - \mathbf{R}/2|} - A\lambda(R) + B\frac{e^{-\lambda(R)R}}{R}, \tag{7.140}$$

$$\frac{1}{a} - \frac{1}{|\mathbf{r} + \mathbf{R}/2|} \propto \frac{B}{|\mathbf{r} + \mathbf{R}/2|} - B\lambda(R) + A\frac{e^{-\lambda(R)R}}{R}. \tag{7.141}$$

Clearly, the proportionality constants in equations (7.140) and (7.141) are, respectively, $1/A$ and $1/B$, since the singular terms there must be cancelled out. Hence, we are left with the following system of equations (we set $A \equiv 1$ without loss of generality)

$$\frac{1}{a} - \lambda(R) + B\frac{e^{-\lambda(R)R}}{R} = 0, \tag{7.142}$$

$$\left(\frac{1}{a} - \lambda(R)\right)B + \frac{e^{-\lambda(R)R}}{R} = 0. \tag{7.143}$$

There are two solutions to equations (7.142) and (7.143), namely $B = \pm 1$. Since we are interested mostly in the unitary limit, we set $1/a = 0$, and the two possible solutions for $\lambda(R)$, which we call $\lambda_\pm(R)$, are obtained by setting $B = \pm 1$ in equation (7.142), for instance. Redefining $\lambda(R)_\pm R \equiv \gamma_\pm$, the equation simplifies to

$$\gamma_\pm = \pm e^{-\gamma_\pm} \Rightarrow \gamma_+ = 0.567\,143\,290\ldots, \tag{7.144}$$

which is, of course, independent of R, while there are infinitely many non-real solutions for γ_-. That is, the only bound state requires to take $B = +1$, and we have a linearly decaying $\lambda_+(R) = \gamma_+/R$. The Born–Oppenheimer potential, equation (7.139), at unitarity, takes the simple form

$$\mathcal{E}(R) = -\frac{\hbar^2\gamma_+^2}{2mR^2}. \tag{7.145}$$

We now move on to the second step in the Born–Oppenheimer approximation. That is, we solve the heavy-heavy stationary Schrödinger equation with the two-body interaction given by the Born–Oppenheimer potential in equation (7.145). We restrict ourselves to the s-wave channel once more. After separating the centre-of-mass coordinate, the Schrödinger equation for the relative wave function $\phi(\mathbf{R})$ of the two heavy particles takes the form

$$\left[-\frac{\hbar^2\nabla_{\mathbf{R}}^2}{M} + \mathcal{E}(R)\right]\phi(\mathbf{R}) = E\phi(\mathbf{R}), \tag{7.146}$$

where $\mathcal{E}$ is given by equation (7.145). Working in the s-wave channel, and defining the reduced radial wave function Φ as $\Phi = R\phi$, as well as $E = -\hbar^2 Q^2/M$, and setting $\Gamma = \gamma_+^2/2(m/M)$, the Schrödinger equation becomes

$$\Phi''(R) + \frac{\Gamma}{R^2}\Phi(R) = Q^2\Phi(R). \tag{7.147}$$

At short distances, $R \to 0$, the term on the right-hand-side of equation (7.147) can be neglected, and we have $\Phi(R) \to R^s$ as $R \to 0$. Inserting this into the Schrödinger equation (7.147) with $Q = 0$, we obtain the following solutions, $s_\pm$ for the exponent s,

$$s_\pm = \frac{1}{2} \pm \sqrt{\frac{1}{4} - \Gamma}. \tag{7.148}$$

If $\Gamma < 1/4$, then the exponents $s_\pm$ are real, and the solutions behave as power laws. For $\Gamma > 1/4$, however, the exponents are complex. This represents the appropriate limit, since the Born–Oppenheimer approximation is valid only for $M/m \gg 1$, which implies $\Gamma \gg 1$. We define, in this case,

$$\sigma_0 = \sqrt{\Gamma - \frac{1}{4}}, \tag{7.149}$$

so that the reduced wave function Φ is given by a linear combination of the form

$$\Phi(R) \propto \sqrt{R}[R^{i\sigma_0} + \tilde{B}R^{-i\sigma_0}], \quad R \to 0. \tag{7.150}$$

Just as when we had identical particles, we need to impose an extra boundary condition, which will break the continuous scale invariance down to discrete scale invariance. We choose a nodal condition, such that $\Phi(R_*) = 0$, and if we choose $\tilde{B} = -1$, we have

$$\Phi(R) \propto \sqrt{R}\sin(\sigma_0 \ln(R/R_*)), \quad R \to 0. \tag{7.151}$$

Here, R_* plays the role of the three-body parameter. Now that we have introduced the important length scale R_*, we solve the full Schrödinger equation, equation (7.148) together with the boundary condition (7.151). Defining $\Phi = \sqrt{R}\chi$, $z = iQR$, and $g(z) = \chi(R)$ equation (7.148) becomes

$$z^2 g''(z) + zg'(z) + [z^2 - (i\sigma_0)^2]g(z) = 0, \tag{7.152}$$

which is nothing but Bessel's differential equation. It solutions are given by

$$g_\pm(z) = J_{\pm i\sigma_0}(z), \tag{7.153}$$

so that the reduced wave function takes the form

$$\Phi(R) = \sqrt{R}[J_{i\sigma_0}(iQR) + \bar{B}Y_{i\sigma_0}(iQR)], \tag{7.154}$$

with $\bar{B}$ a constant, and where J_ν and Y_ν are, respectively, Bessel functions of the first and second kind. For $\Phi(R)$ in equation (7.154) to be well defined as $R \to \infty$, that is, to calculate $\bar{B}$, we invoke the asymptotic limit of the Bessel functions J_ν and Y_ν. This is well known, and takes the form

$$J_\nu(z) \to \sqrt{\frac{2}{\pi z}}\cos\left(z - \frac{\nu\pi}{2} - \frac{\pi}{4}\right), \tag{7.155}$$

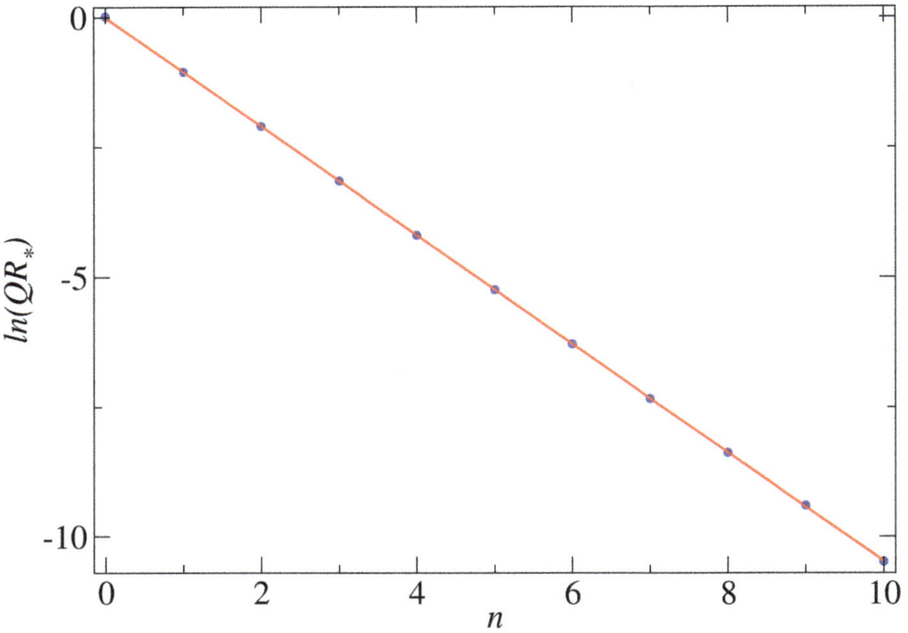

Figure 7.5. Numerically calculated $\ln(Q_n R_*)$ (blue dots) for $s_0 = 3$, corresponding to $\Gamma = 37/4$ (a mass ratio of $m/M \approx 0.017$), as function of the state number, compared with the prediction $\ln(QR_*) = \ln(Q_0 R_*) - \pi n/s_0$ (red line).

$$Y_\nu(z) \to \sqrt{\frac{2}{\pi z}} \sin\left(z - \frac{\nu\pi}{2} - \frac{\pi}{4}\right). \tag{7.156}$$

Therefore, we must have $\bar{B} = i$, so that $\Phi(R) \propto \exp(-QR)$ as $R \to \infty$. We then obtain the ground state energy as the largest value, Q_0, of Q that makes $\Phi(R_*) = 0$, which is obtained numerically. The excited states are also obtained numerically, and they satisfy very well the correct Efimov scaling, that is,

$$Q_{n+1} = Q_n e^{-\pi/s_0}, \quad n = 1, 2, \ldots. \tag{7.157}$$

This is clearly observed in figure 7.5, where we plot the logarithm of the binding momenta from the ground state until the 10th excited state.

7.3 Exactly solvable three-body problems

We have already seen a fair selection of exactly solvable three-body problems. Those that exhibit the Efimov effect and can be solved exactly feature a strong symmetry at the classical level—continuous scale invariance—while at the quantum level this is broken down to discrete scale invariance. This is the reason why they may be solved essentially by hand. The one-dimensional three-boson problem with LO EFT interactions was also solved exactly (for the ground state with attraction) via the Faddeev equations in the momentum representation. However, the solution there requires some non-trivial manipulation and integration. Essentially, thus, we have solved some rather important three-body problems, but we have no way to systematize these solutions so we can extend them to other non-trivial and interesting problems. In this section, our goal is to present some systematics that will allow us to not only solve certain three-body problems exactly, but also construct other exactly solvable models. Unfortunately, this can be done only in one spatial dimension (with some exceptions), but gladly one-dimensional systems are experimentally relevant in many different fields and some of the exactly solvable models are actually the main tool for reduced-dimensional ensembles of ultracold atomic gases.

7.3.1 Product ground states: supersymmetric quantum mechanics

We begin with the simplest of all methods, which lies underneath the so-called supersymmetric quantum mechanics (SuSyQM). The full machinery of SuSyQM is beyond the scope of this book, but we refer the interested reader to the excellent review by Cooper and coworkers on the subject [17]. The main idea, at least from a practical standpoint, is to solve for the ground state of a quantum system that has the form of a product. For three particles, this means that the ground state in the position representation, $\psi_0(\mathbf{r}_1, \mathbf{r}_2, \mathbf{r}_3)$, takes the form

$$\psi_0(\mathbf{r}_1, \mathbf{r}_2, \mathbf{r}_3) = \phi(\mathbf{r}_1, \mathbf{r}_2)\phi(\mathbf{r}_1, \mathbf{r}_3)\phi(\mathbf{r}_2, \mathbf{r}_3), \tag{7.158}$$

where ϕ is a two-body function such that ψ_0 is in the Hilbert space (except for, perhaps, a global centre-of-mass plane-wave that separates). Clearly, equation (7.158) implies that our system is bosonic if $\phi(\mathbf{r}_i, \mathbf{r}_j).= \phi(\mathbf{r}_j, \mathbf{r}_i)$. We will stick to bosonic states in this section.

To begin, let us consider a symmetric function $\phi(\mathbf{r}_i, \mathbf{r}_j) = \phi(\mathbf{r}_{ij})$, where $\mathbf{r}_{ij} = \mathbf{r}_i - \mathbf{r}_j$. We impose the constraint on $\phi(\mathbf{r}_{ij})$ that it is normalizable in the relative distance, that is

$$\int d\mathbf{r}_{ij}|\phi(\mathbf{r}_{ij})|^2 < \infty. \tag{7.159}$$

The next constraint is that it must be an eigenstate of some non-relativistic two-body, relative Hamiltonian H, that is, there exists H such that

$$H\phi = -\frac{\hbar^2}{m}\nabla^2\phi(\mathbf{r}_{ij}) + V(\mathbf{r}_{ij})\phi(\mathbf{r}_{ij}) = E\phi(\mathbf{r}_{ij}). \tag{7.160}$$

If we know ϕ and E, then we immediately know the Hamiltonian, since,

$$V(\mathbf{r}_{ij}) = \frac{1}{\phi(\mathbf{r}_{ij})}\left[E + \frac{\hbar^2}{m}\nabla^2\right]\phi(\mathbf{r}_{ij}), \tag{7.161}$$

which is valid for *nodeless* wave functions, that is, we should be looking for the ground state if we do not want to run into trouble. However, we usually do know the potential, and from it we can calculate ϕ, for the two-body problem, without many issues (the two-body problem, as the reader should know by now, is quite manageable whether analytically or numerically). What we would like to know, however, is, given V, and therefore ϕ, whether our three-body Hamiltonian can have a ground state of the type (7.158). To do this, we write down the following auxiliary two-body Hamiltonian H_{SuSy},

$$H_{\text{SuSy}} \equiv \frac{\hbar^2}{2m} A^{\dagger} A, \tag{7.162}$$

where the operators A (and $A^{\dagger}$ as its adjoint) are given by

$$A = \hat{r}_0 \cdot \nabla + W(\mathbf{r}_{ij}). \tag{7.163}$$

Above, W will be called the *superpotential*, while $\hat{\mathbf{r}}_0$ is a unit vector, which is required so that H_{SuSy} is a scalar and not a vector. Now we see why one-dimensional systems are strongly preferred: we only obtain a nice non-relativistic auxiliary Hamiltonian in one dimension, where the operator A is simply

$$A = \frac{\partial}{\partial x_{ij}} + W(x_{ij}). \tag{7.164}$$

From now on, we consider only one-dimensional systems. The auxiliary Hamiltonian, equation (7.162), takes the following expanded form

$$H_{\text{SuSy}} = \frac{\hbar^2}{m}\left[-\frac{\partial^2}{\partial x_{ij}^2} - W'(x_{ij}) + \left[W(x_{ij})\right]^2\right]. \tag{7.165}$$

Therefore, the potential, V_{SuSy}, present in the auxiliary Hamiltonian is given by

$$V_{\text{SuSy}}(x_{ij}) = \frac{\hbar^2}{m}\left\{\left[W(x_{ij})\right]^2 - W'(x_{ij})\right\}. \tag{7.166}$$

Before directly trying to identify V_{SuSy} with the original potential, we look at the auxiliary Hamiltonian in the form (7.162), where it is clear that H_{SuSy} is a semi-positive operator, and therefore its ground state energy is positive or zero. Therefore, if we can find a normalizable ϕ such that $A\phi = 0$, then the ground state energy of H_{SuSy} is zero. Since an arbitrary Hamiltonian can have a non-zero ground state energy, say E_0, then if the annihilated function ϕ is normalizable,

$$V_{\text{SuSy}} = V - E_0. \tag{7.167}$$

The simplification here is essentially non-existent so far, for two bodies, unless we propose a form for W. If however, we know the superpotential W, then annihilating the ground state reduces to a first-order differential equation $A\phi = 0$, which explicitly reads

$$\phi'(x_{ij}) + W(x_{ij})\phi(x_{ij}) = 0 \Rightarrow \phi(x_{ij}) = C\exp\left[-\int dx_{ij} W(x_{ij})\right], \tag{7.168}$$

where C is a constant. Note that for ϕ to be normalizable, necessarily $W(x)$ is an antisymmetric function. Moreover, W is allowed to be a distribution. The simplest example of it is given by the choice

$$W(x_{ij}) = c_0 \,\mathrm{sgn}(x_{ij}). \tag{7.169}$$

The ground state wave function associated with the above superpotential is

$$\phi(x_{ij}) \propto e^{-c_0|x_{ij}|}, \tag{7.170}$$

which is normalizable if $c_0 > 0$. The aware reader must have noticed that equation (7.170) is exactly the ground state of a one-dimensional two-body system with attractive LO EFT interactions, that is, the potential V is a Dirac delta function with negative coupling constant g_0. We see that this is the case, too, by calculating V_{SuSy} explicitly from equation (7.166), which gives

$$V_{\mathrm{SuSy}} = -\frac{2\hbar^2 c_0}{m}\delta(x_{ij}) + \frac{\hbar^2 c_0^2}{m}. \tag{7.171}$$

By comparison with equation (7.167) and the usual LO EFT in one dimension, we have

$$g_0 = -\frac{2\hbar^2 c_0}{m}, \qquad E_0 = -\frac{m g_0^2}{4\hbar^2}, \tag{7.172}$$

which is exactly what we expected.

We are now going to turn the above trivial walkthrough into an interesting and useful method, by considering three bosons in one dimension. We build our auxiliary Hamiltonian now in the following form

$$H_{\mathrm{SuSy}} = \frac{\hbar^2}{2m}\sum_{i=1}^{3} A_i^\dagger A_i, \tag{7.173}$$

where A_i depends on the coordinates of *all three particles*, but only includes the partial derivative with respect to x_i. That is, we consider

$$A_i = \frac{\partial}{\partial x_i} + W_i(x_1, x_2, x_3). \tag{7.174}$$

The auxiliary Hamiltonian in equation (7.172) is still semi-positive, since it is the sum of three semi-positive operators. Hence, if there exists a normalizable $\phi(x_1, x_2, x_3)$ such that

$$A_i\phi = 0, \quad i = 1, 2, 3, \tag{7.175}$$

then ϕ is the ground state of H_{Susy} in equation (7.173) and its ground state energy vanishes. In the three-body case, however, ϕ must satisfy a system of three partial differential equations (PDEs). Since the PDEs are first order, they are not difficult to solve. We have

$$\partial_{x_i}\phi(x_i, x_j, x_\ell) + W_i(x_1, x_2, x_3)\phi(x_i, x_j, x_\ell) = 0, \tag{7.176}$$

where the choice of labels $(ij\ell)$ and (123) is *deliberate and necessary*. Equation (7.176) is solved by

$$\phi(x_i, x_j, x_\ell) = F_i(x_j, x_\ell)\exp\left[-\int dx_i W_i(x_1, x_2, x_3)\right]. \tag{7.177}$$

Above, F_i is a function that does not depend on x_i, but can surely depend on x_j and x_ℓ, as we know from the standard theory of partial differential equations. Since ϕ must be bosonic and normalisable, we must analyse the symmetry of F_i and W_i.

We begin by proposing a particular form for W_i, consisting only of two-body terms, as

$$W_i(x_i, x_j, x_\ell) = W(x_i - x_j) + W(x_i - x_\ell), \tag{7.178}$$

where W is some function. For ϕ in equation (7.177) to be the same for all $i = 1, 2, 3$, W must be antisymmetric. In that case, calling

$$B(x_i - x_j) = \exp\left[-\int dx_i W(x_i - x_j)\right], \tag{7.179}$$

we immediately have

$$F_i(x_j, x_\ell) = B(x_j - x_\ell), \tag{7.180}$$

and the solution ϕ takes the form

$$\phi(x_1, x_2, x_3) = B(x_1 - x_2)B(x_1 - x_3)B(x_2 - x_3), \tag{7.181}$$

which is exactly of the product form (7.158).

Another possible choice is a set of W_i's that is not separable into two-body pieces, i.e., not of the form in equation (7.178). A simple ansatz for W_i that can satisfy normalizability and symmetry, and is similar in spirit to equation (7.178), is the choice

$$W_i(x_i, x_j, x_\ell) = W^{(3)}\left(x_i - \frac{x_j + x_\ell}{2}\right) + W^{(2)}(x_i - x_j) + W^{(2)}(x_i - x_\ell). \tag{7.182}$$

Clearly, the above superpotential contains a two-body term identical to that in equation (7.178), in addition to a three-body term that describes interactions between particle i and the centre-of-mass of the $j\ell$ subsystem. Denoting $X_{j\ell} = (x_j + x_\ell)/2$, we see that equation (7.182) has the same symmetry properties as equation (7.178) if we choose $W^{(3)}$ to be antisymmetric with respect to the exchange of x_i and $X_{j\ell}$.

There are certainly more generic forms for W_i that lead to admissible annihilated states. But we shall not pursue this any further since, in any case, this is an *open problem* at the time of writing.

We now claim (see problem 4) that some of the classic exactly solvable three-body problems in one dimension can be treated within the formalism above. These all have in common that W_i is of the form of equation (7.178), and with the properties for W we have discussed. Moreover, their Hamiltonian (7.173) does not feature any three-body interaction. That is, the three-body cross-terms generated by multiplying $A_i^\dagger$ and A_i add up to constants and two-body terms only. This, in fact, leads to a (highly non-trivial) equation for W. To see this, we simply write down H_{SuSy} with W_i in equation (7.178). We obtain

$$H_{\text{SuSy}} = -\frac{\hbar^2}{2m}\sum_i \frac{\partial}{\partial x_i} + V_{\text{Susy}}, \tag{7.183}$$

where the potential is given by

$$\begin{aligned} \frac{2m}{\hbar^2}V_{\text{Susy}} = & \sum_{i\neq k}\left\{-\frac{\partial}{\partial x_i}W^{(2)}(x_i - x_k) + [W^{(2)}(x_i - x_k)]^2\right\} \\ & + 2\sum_i W^{(2)}(x_i - x_j)W^{(2)}(x_i - x_\ell). \end{aligned} \tag{7.184}$$

The terms in the bottom line of equation (7.184) correspond to three-body interactions unless, as we have noted above, they add up to a constant or two-body terms only. If we develop the terms explicitly, and equate them to a two-body function $\sum_{ij} f(x_{ij})$, we have

$$W^{(2)}(x_{12})W^{(2)}(x_{13}) + W^{(2)}(x_{21})W^{(2)}(x_{23}) + W^{(2)}(x_{31})W^{(2)}(x_{32}) = \sum_{i\neq j} f(x_{ij}). \tag{7.185}$$

Solutions to the above equation are actually known in general, but we will simply point out the most important ones. Three superpotentials that are solutions to equation (7.185) are given by

$$W^{(2)}(x) = \alpha x, \tag{7.186}$$

$$W^{(2)}(x) = \beta\ \text{sgn}(x), \tag{7.187}$$

$$W^{(2)}(x) = \frac{\gamma}{x}. \tag{7.188}$$

We urge the reader (see problem 4) to show that the above superpotentials do satisfy equation (7.185) and, therefore, yield Hamiltonians that include only two-body potentials. These in fact are the well known Calogero, attractive Lieb–Liniger (or McGuire), and harmonic interactions which, as it turns out, are also exactly solvable.

7.3.2 Integrable systems

There is a technique that was introduced by H A Bethe a long time ago in the context of spin chains, commonly known as the Bethe ansatz, that solves exactly a number of important and interesting few- and many-body interacting problems in one spatial dimension. The ansatz on its own only gives the exact functional form of the wave functions, but the name 'Bethe ansatz' is used nowadays to refer to a collection of techniques, some more involved than others, that use the ansatz of Bethe to solve interacting problems in one dimension exactly.

Before studying the Bethe ansatz for the two- and three-body problems, we discuss briefly the classical-mechanical notion of integrability, which is tightly linked to exact solvability (although these are not the same concept). In the quantum-mechanical part of this section, we will actually talk about integrability and solvability interchangeably. In classical mechanics, a one-dimensional system is said to be integrable, roughly speaking, when the number of conserved quantities equals the number of particles in the system. This implies that every one-dimensional, non-relativistic, two-body system of identical particles with translational invariance is integrable, since they conserve the Hamiltonian and the total momentum. Note that we still need to solve Newton's equations for the relative motion of the pair, which might need numerical methods. But this is the simplest possible problem in classical mechanics and if we can reduce a three-body problem to another problem with the complexity of a one-body system, we can consider it exactly solvable (although, strictly speaking, it is only integrable unless we can solve the problem by hand). In classical mechanics, there are a number of techniques designed specifically to attack integrable models and to decide upon the integrability of a particular system. One such technique makes use of the so-called Lax pairs. Although mathematically it may appear, as we will see, that classical integrability has *a priori* not much to do with quantum integrability, it is true that classical integrable models tend to be integrable in the quantum case, too. Moreover, we will show that classical integrability via Lax pairs is highly connected with models exhibiting SuSyQM, which we saw in section 7.3.1.

We begin with a short reminder of classical mechanics in the Hamiltonian formalism, so that we can introduce the concepts and notation necessary in order to define integrability in the classical sense. We will follow the excellent book on integrable systems by Calogero [18]. Our system consists of N non-relativistic particles in one dimension whose dynamics is determined by a time-independent Hamiltonian H that depends on the canonical coordinates $q \equiv (q_1, \dots, q_N)$, and their canonically conjugate momenta $p = (p_1, \dots, p_N)$. We write the Hamiltonian $H = H(q, p)$, and the dynamics is given by Hamilton's equations of motion,

$$\dot{q}_i = \frac{\partial H}{\partial p_i}(q, p), \tag{7.189}$$

$$\dot{p}_i = -\frac{\partial H}{\partial q_i}(q, p). \tag{7.190}$$

We have used Newton's dot notation to indicate derivative with respect to time, that is, $\dot{A} \equiv dA/dt$. Given the set of canonical coordinates and momenta, and two classical observables $A = A(q, p)$ and $B = B(q, p)$, we define their Poisson bracket with respect to the canonical coordinates (q, p) as

$$\{A, B\}_{q,p} \equiv \sum_{i=1}^{N}\left[\frac{\partial A}{\partial q_i}\frac{\partial B}{\partial p_i} - \frac{\partial A}{\partial p_i}\frac{\partial B}{\partial q_i}\right]. \tag{7.191}$$

Whenever there is no room for confusion, we shall denote $\{A, B\}_{q,p} = \{A, B\}$, without reference to the set of coordinates and momenta. It is straightforward to show, using Hamilton's equations (7.189) and (7.190), that the time evolution of an observable $A = A(q, p)$ is given by

$$\dot{A} = \{A, H\}. \tag{7.192}$$

Therefore, if the Poisson bracket of an observable and the Hamiltonian vanishes, i.e., if $\{A, H\} = 0$, then by equation (7.192) A is a constant of motion, that is, $\dot{A} = 0$.

We define now canonical transformations from a set (q, p) to another set (Q, P). We say that the transformation taking (q, p) into (Q, P) is canonical if

$$\{Q_i, P_j\}_{q,p} = \delta_{ij}, \tag{7.193}$$

$$\{Q_i, Q_j\}_{q,p} = \{P_i, P_j\}_{q,p} = 0. \tag{7.194}$$

Canonical transformations are special because the Hamiltonian, when written as a function of (Q, P), produces the same equations of motion as the original Hamiltonian in terms of (q, p). That is, if $\mathcal{H}(Q, P) = H(q, p)$ is the Hamiltonian, then if (Q, P) satisfy equations (7.193) and (7.194) (i.e., if the transformation is canonical), we have

$$\dot{Q}_i = \frac{\partial \mathcal{H}}{\partial P_i}(Q, P), \tag{7.195}$$

$$\dot{P}_i = -\frac{\partial \mathcal{H}}{\partial Q_i}(Q, P). \tag{7.196}$$

The proof of the above is very simple.

There are special kinds of canonical coordinates, called action-angle variables, that we introduce now. Let us assume that we have the canonical coordinates and momenta (Q, P), and the Hamiltonian $\mathcal{H}(Q, P)$. Imagine that $\mathcal{H}$ does not depend on a particular canonical coordinate, which we may choose to be Q_{i_0}. This means that

$$\dot{P}_{i_0} = -\frac{\partial \mathcal{H}}{\partial Q_{i_0}} = 0. \tag{7.197}$$

That is, if the Hamiltonian does not depend on Q_{i_0}, then its canonically conjugate momentum is a constant of motion. Of course, if $\mathcal{H}(Q, P)$ is independent of all canonical coordinates, that is, $\mathcal{H} = \mathcal{H}(Q)$, then all canonically conjugate momenta are constants of motion. If this is the case, then the canonical coordinates move as those of free particles with constant speeds. Such set (Q, P) receives the name of action-angle variables, and they play an important role in the theory of integrable systems.

We are now ready to define classical integrability. Assume that an N-body classical system in one dimension possesses N constants of motion $\{P_i(q, p)\}_{i=1}^{N}$, which are single-valued and independent of each other. Assume that their mutual Poisson brackets vanish, that is,

$$\{P_i, P_j\} = 0 \quad \forall\, i, j. \tag{7.198}$$

Then the system is called *Liouville Integrable*. The constants of motion, regarded as canonically conjugate momenta P (the notation was on purpose), together with the set of canonical coordinates Q, that P are conjugate to, if these can be identified, represent a set (Q, P) of action-angle variables.

We introduce now the concept of a Lax pair, which helps identify and construct integrable models as well as their constants of motion. We say that two (implicitly time-dependent) matrices L and M constitute a Lax pair if the time evolution is given by the following equation (the Lax equation)

$$\dot{L} = [L, M]. \tag{7.199}$$

For the Lax pair to be useful in classical mechanics, equation (7.199) must be compatible with Newtonian time evolution (Newtonian dynamics). This is because if $L = L(q, p)$ and $M = M(q, p)$, equation (7.199) has N^2 independent equations and only $2N$ variables, N canonical coordinates $\{q_j\}_j$ and N canonically conjugate momenta $\{p_j\}_j$. The importance of the Lax equation, equation (7.199), is the fact that it implies that there are N constants of motion. These are given by the eigenvalues of L, which turn out to be independent of time, as we will see shortly. For this to be the case, obviously we need the assumption that L is diagonalizable. We call its eigenvalues $\lambda^{(j)}$, so that there are vectors $|v^{(j)}\rangle$ and $|u^{(j)}\rangle$ satisfying

$$L|v^{(j)}\rangle = \lambda^{(j)}|v^{(j)}\rangle, \tag{7.200}$$

$$\langle u^{(j)}|L = \lambda^{(j)}\langle u^{(j)}|. \tag{7.201}$$

Above, and as long as we are dealing with integrability in classical systems, $\langle u^{(j)}|$ is the transpose of $|u^{(j)}\rangle$, and the scalar product between $u^{(j)}$ and $v^{(j)}$ is defined as

$$\langle u^{(j)}|v^{(k)}\rangle = \sum_{n=1}^{N} u_n^{(j)} v_n^{(k)}, \tag{7.202}$$

that is, we *do not take the complex conjugate of the elements of* $|u^{(j)}\rangle$ when taking scalar products. Also, we have

$$\langle u^{(j)}|Av^{(k)}\rangle = \langle A^{\mathrm{T}}u^{(j)}|v^{(k)}\rangle, \tag{7.203}$$

where A is a matrix, not necessarily real. Whenever there is no confusion, we will abuse the notation for Hermitian matrices in quantum mechanics and write

$$\langle u^{(j)}|A|v^{(k)}\rangle = \langle u^{(j)}|Av^{(k)}\rangle, \tag{7.204}$$

to denote the untransposed matrix applying on the right vector. A direct consequence of diagonalizability is the fact that the scalar product between right and left eigenvectors corresponding to different eigenvalues $\lambda(j) \neq \lambda^{(k)}$ vanishes, that is,

$$\text{if} \quad \lambda^{(j)} \neq \lambda^{(k)} \Rightarrow \langle u^{(j)}|v^{(k)}\rangle = 0. \tag{7.205}$$

The proof of this is straightforward, simply calculate

$$\langle u^{(j)}|L|v^{(k)}\rangle = \lambda^{(j)}\langle u^{(j)}|v^{(k)}\rangle = \lambda^{(k)}\langle u^{(j)}|v^{(k)}\rangle \Rightarrow \langle u^{(j)}|v^{(k)}\rangle = 0. \tag{7.206}$$

We can (and will) choose the normalization of the left and right eigenvectors of the Lax matrix L so that they are orthonormal, i.e.,

$$\langle u^{(j)}|v^{(k)}\rangle = \delta_{jk}. \tag{7.207}$$

Our second assumption is that all eigenvalues are different from each other. This implies equation (7.207) with our choice of normalization. In turn, equation (7.207) implies the completeness relation

$$\sum_{j=1}^{N} u_n^{(j)} v_m^{(k)} = \delta_{n,m}, \tag{7.208}$$

which is straightforward to show, and the reader is encouraged to prove on their own.

We are now fully ready to show that the eigenvalues of the Lax matrix L, under the assumptions we have laid out, are time independent, that is, we are going to prove that

$$\dot{\lambda}^{(j)} = 0, \quad j = 1, 2, \dots, N. \tag{7.209}$$

The proof is quite simple. Using equations (7.200) and (7.207), it follows immediately that

$$\lambda^{(j)} = \langle u^{(j)}|L|v^{(j)}\rangle. \tag{7.210}$$

To simplify the notation a little, we suppress the superscript (j) from the eigenvalue and eigenvectors, and fix it to (j). Differentiating the relation above with respect to time, we obtain

$$\dot{\lambda} = \langle \dot{u}|L|v\rangle + \langle u|\dot{L}|v\rangle + \langle u|L|\dot{v}\rangle. \tag{7.211}$$

Using the Lax evolution equation for L, equation (7.199), in equation (7.211), and the fact that u and v are eigenvectors with eigenvalue λ, we obtain

$$\dot{\lambda} = \lambda[\langle \dot{u}|v\rangle + \langle u|\dot{v}\rangle] + \langle u|[L, M]|v\rangle. \tag{7.212}$$

Moreover, due to equation (7.207), we have

$$0 = \frac{d}{dt}\langle u|v\rangle = \langle \dot{u}|v\rangle + \langle u|\dot{v}\rangle. \tag{7.213}$$

Using the above result in equation (7.212), expanding the commutator and using equations (7.200) and (7.201), we have

$$\dot{\lambda} = \langle u|[L, M]|v\rangle = \lambda\langle u|M|v\rangle - \lambda\langle u|M|v\rangle = 0, \tag{7.214}$$

as we wanted to prove.

What remains to be done now, and can be non-trivial, is to find particular forms of the Lax pair L and M whose evolution (7.199) corresponds to Hamiltonian evolution of an N-particle problem. If we can find one such Lax pair, then the eigenvalues of L will be constants of motion. Moreover, the system will be integrable if the constants of motion are non-trivial (for instance, that they are not just N powers of the Hamiltonian), functionally independent of each other (so that they correspond, in the sense of Liouville integrability, with N constants of motion), and that their Poisson brackets all vanish. We do this in what follows for the simplest of models.

We define three functions of the (relative) canonical coordinates $q_{nm} = q_n - q_m$, which we call α, β and γ. For simplicity in the notation, we shall also write f_{nm} to refer to $f(q_{nm})$, when there is no possible confusion. We also call $p_{nm} = p_n - p_m$ (twice) the relative momentum between particles labelled n and m. The Lax pair is then chosen as

$$L_{n,m} = p_n\delta_{n,m} + \alpha_{n,m}(1 - \delta_{n,m}), \tag{7.215}$$

$$M_{n,m} = \delta_{n,m}\sum_{\ell=1}^{N}{}'\beta_{n\ell} + \gamma_{n,m}(1 - \delta_{n,m}). \tag{7.216}$$

Above, the primed sum means that the sum is performed over $\ell \neq n$. We insert equations (7.215) and (7.216) into the Lax evolution equation. After tedious but straightforward manipulations, we obtain

$$\dot{p}_n = \sum_{\ell=1}^{N}{}'[\alpha_{n\ell}\gamma_{n\ell} - \gamma_{\ell n}\alpha_{\ell n}], \tag{7.217}$$

as well as

$$\dot{q}_{nm}\partial_{q_n}\alpha_{nm} = p_{nm}\gamma_{nm} + \alpha_{nm}\left(\sum_{\ell}{}'\beta_{m\ell} - \sum_{\ell}{}'\beta_{n\ell}\right) + \sum_{\ell\neq n,m}(\alpha_{n\ell}\gamma_{\ell m} - \alpha_{\ell m}\gamma_{n\ell}). \tag{7.218}$$

Since we require that equations (7.217) and (7.218) correspond to Hamiltonian evolution, we have $p_{nm} = \dot{q}_{nm}$ as well as $p_n = \dot{q}_n$. It is easy to see that a solution to equation (7.218) is obtained if we set $\gamma_{nm} = \partial_{q_n}\alpha_{nm}$, $\beta_{m\ell} = \beta_{\ell m}$, and

$$\alpha_{nm}\left[\beta_{m\ell} - \beta_{n\ell}\right] + \alpha_{n\ell}\gamma_{\ell m} - \alpha_{\ell m}\gamma_{n\ell} = 0. \tag{7.219}$$

Inserting all of the above into equation (7.217), we have

$$\dot{p}_n = \sum_{\ell}{}' \left[\alpha_{n\ell}\partial_{q_\ell}\alpha_{\ell n} - \alpha_{\ell n}\partial_{q_n}\alpha_{n\ell}\right]. \tag{7.220}$$

The above equation of motion does correspond to Hamiltonian evolution, with the interaction potential V between particles n and m given by

$$V(q_n - q_m) = \alpha(q_n - q_m)\alpha(q_m - q_n). \tag{7.221}$$

Therefore, we have succeeded in finding an acceptable example of a Lax pair that gives rise to Hamiltonian evolution for an interacting N-body classical system.

To motivate the choice of Lax matrices, as well as to show, using the Lax pair technique, that two-body one-dimensional systems are integrable, we take a look at the simple two-body problem. The Lax matrix L is given in this case by

$$L = \begin{pmatrix} p_1 & \alpha_{12} \\ \alpha_{21} & p_2 \end{pmatrix}. \tag{7.222}$$

The other Lax matrix M takes also a simple form, but we do not write it here. The eigenvalues $\lambda_\pm$ of the Lax matrix in equation (7.222) are given by

$$\lambda_\pm = \frac{1}{2}\left[p_1 + p_2 \pm \sqrt{(p_1 + p_2)^2 - 4p_1p_2 + 4\alpha_{12}\alpha_{21}}\right]. \tag{7.223}$$

Instead of the eigenvalues, we may take a different set which may consist of its trace,

$$\mathrm{Tr}\, L = p_1 + p_2 \equiv P, \tag{7.224}$$

which is just the total momentum, indeed a conserved quantity, and the difference $d = \lambda_+ - \lambda_-$, which is given by

$$d = \sqrt{P^2 - 4(p_1p_2 - \alpha_{12}\alpha_{21})}. \tag{7.225}$$

Since d is a conserved quantity, so is its square. And since P is conserved, we can add it or subtract it at will, so the following quantity is also conserved,

$$d' = p_1p_2 - \alpha_{12}\alpha_{21}. \tag{7.226}$$

It turns out that d' above is actually a symmetric invariant of L. The N symmetric invariants of a $N \times N$ matrix L J_j are defined via

$$\det(L - \gamma 1) = \gamma^N + \sum_{j=1}^{N} \gamma^{j-1} J_j. \tag{7.227}$$

In the two-dimensional case, $J_1 = d'$ in equation (7.226), while $J_2 = -P$. We complete d' so that we can relate it to another conserved quantity, namely the Hamiltonian. Take the relative momentum $p = (p_1 - p_2)/2$. The relative Hamiltonian H_r, using equation (7.221) is

$$H_r = p^2 + \alpha_{12}\alpha_{21} = \frac{1}{4}\left[p_1^2 + p_2^2 - 2p_1p_2\right] + \alpha_{12}\alpha_{21} = -(p_1p_2 - \alpha_{12}\alpha_{21}) + \frac{P^2}{4} = -J_1 + \frac{J_2^2}{4}. \tag{7.228}$$

Therefore, the relative Hamiltonian is a constant of motion, and so is the total Hamiltonian.

After this classical interlude, we are ready to take on quantum-mechanical integrability, at least in some situations. Following what we have already discussed, at quite some length, in the classical case, by introducing the quantum version of the Lax equation, equation (7.199). We consider a square matrix arrangement of linear operators (not necessarily Hermitian or self-adjoint), that is, we consider N^2 operators $\{L_{ij}\}_{i,j=1}^{N}$, and the matrix arrangement is called L. The quantum Lax equation consists of choosing a quantum Lax pair, L and M, so that

$$[H, L_{jk}] = \sum_{m=1}^{N} \left[M_{jm} L_{mk} - L_{jm} M_{mk} \right]. \tag{7.229}$$

We note that the operators L_{jk} and M_{jk} are time independent, that is, they are operators in the Schrödinger picture. The left-hand-side of equation (7.229) would actually satisfy

$$\dot{L}_{jk}(t) = [H, L_{jk}(t)], \tag{7.230}$$

were L_{jk} written in the Heisenberg picture instead (which is what we mean by $L_{jk}(t)$). However, and anticipating the results we will obtain, this will not be relevant since we will find constants of motion, that is, operators which are time independent in the Heisenberg picture, therefore coinciding with their Schrödinger picture. Nonetheless, the Heisenberg equations of motion should be taken only as inspiration for requiring equation (7.229) to hold. In analogy with the classical case, we shall consider

$$L_{jk} = \delta_{jk} p_j + i(1 - \delta_{jk})\alpha_{jk}, \tag{7.231}$$

$$M_{jk} = \delta_{jk} \sum_{m=1}^{N'} \gamma_{jm} + (1 - \delta_{jk})\beta_{jk}. \tag{7.232}$$

Above, the only difference with respect to the classical case is the fact that p_j and x_j do not commute, but instead satisfy

$$[x_j, p_k] = i\hbar\delta_{jk}. \tag{7.233}$$

The above canonical commutation relation also implies that, given a function $f(x)$,

$$[f(x), p] = i\hbar \frac{df}{dx}, \tag{7.234}$$

a fact that we will use in just a moment.

We begin with the left-hand-side of the diagonal terms of the quantum Lax equation, equation (7.229). Using equation (7.234), we have

$$[H, L_{jj}] = [H, p_j] = i \sum_{m=1}^{N'} \frac{dv_{jm}}{dx_{jm}}, \tag{7.235}$$

where $v_{jm} = V(x_j - x_m)$ and $x_{jm} = x_j - x_m$. It is at this spot that we impose the Hamiltonian being of the form

$$H = \sum_{j=1}^{N} \frac{p_j^2}{2m} + \sum_{j<m=1}^{N} V(x_j - x_m). \tag{7.236}$$

Imposing this is the quantum-mechanical equivalent to imposing Newtonian equations of motion in the classical case. The right-hand-side of equation (7.229) for the diagonal terms is easily obtained as well, and we have

$$i\sum_{m=1}^{N'} \frac{dv_{jm}}{dx_{jm}} = \left[\sum_{k=1}^{N'} \gamma_{jk}, p_j\right] + i\sum_{m=1}^{N'} \left(\beta_{jm}\alpha_{mj} - \alpha_{jm}\beta_{mj}\right). \tag{7.237}$$

We now take a look at the off-diagonal terms of the quantum Lax equation. The left-hand-side of equation (7.229), for $j \neq k$, gives us

$$[H, L_{jk}] = \left[\sum_{\ell=1}^{N} \frac{p_\ell^2}{2} + V, i\alpha_{jk}\right], \tag{7.238}$$

where V is shorthand for the sum of all pairwise potentials. Since $[V, \alpha_{jk}] = 0$, and $\alpha_{jk} = \alpha(x_j - x_k)$, equation (7.238) becomes

$$[H, L_{jk}] = i[p_j^2/2 + p_k^2/2, \alpha_{jk}] = p_j\frac{d\alpha_{jk}}{dx_{jk}} - \frac{d\alpha_{jk}}{dx_{jk}}p_k. \tag{7.239}$$

The right-hand-side is more tedious to simplify, but is equally straightforward. For $j \neq k$, we obtain

$$\begin{aligned}\sum_{m=1}^{N}\left(M_{jm}L_{mk} - L_{jm}M_{mk}\right) &= \beta_{jk}p_k - p_j\beta_{jk}\\ &+ i\sum_{m(\neq j,k)}\left\{\alpha_{jk}[\gamma_{jm} - \gamma_{km}] + \beta_{jm}\alpha_{mk} - \alpha_{jm}\beta_{mk}\right\}.\end{aligned} \tag{7.240}$$

From all of these tedious calculations, the quantum Lax equation implies the following relations between the interaction potential, α, β and γ,

$$\frac{d\alpha}{dx}(x) = -\beta(x), \tag{7.241}$$

$$0 = \alpha(x + y)[\gamma(y) - \gamma(x)] + \beta(y)\alpha(x) - \beta(x)\alpha(y), \tag{7.242}$$

$$\frac{dV}{dx}(x) = \frac{d\gamma}{dx}(x) - 2\beta(x)\alpha(x). \tag{7.243}$$

Introducing equation (7.241) into equation (7.243), we see that

$$\frac{dV}{dx} = \frac{d\gamma}{dx} + 2\frac{d\alpha}{dx}\alpha(x) = \frac{d}{dx}(\gamma + \alpha^2) \Rightarrow V = \gamma + \alpha^2 + C, \tag{7.244}$$

where C is a constant. Relation (7.242), in turn, takes the form

$$\alpha(x + y)[\gamma(y) - \gamma(x)] = \frac{d\alpha}{dy}(y)\alpha(x) - \frac{d\alpha}{dx}(x)\alpha(y). \tag{7.245}$$

A convenient choice for γ (we shall see why in a minute), is to take it as

$$\gamma = -\beta. \tag{7.246}$$

The operators M_{jk} in equation (7.232) simplify to

$$M_{jk} = \delta_{jk}\sum_{m=1}^{N'}\gamma_{jm} - (1 - \delta_{jk})\gamma_{jk}. \tag{7.247}$$

Let us choose a column vector $\vec{\eta}$ and its adjoint $\vec{\eta}^\dagger$ (row vector), with all elements equal to 1, that is, $\eta_j = 1$ $(j = 1, 2, \ldots, N)$. Then

$$(M\vec{\eta})_j = \sum_{k=1}^{N} M_{jk}\eta_k = \sum_{k=1}^{N} M_{jk} = \sum_{m=1}^{N'}\gamma_{jm} - \sum_{k=1}^{N'}\gamma_{jk} = 0 = (\vec{\eta}^\dagger M)_j. \tag{7.248}$$

Therefore, $M\vec{\eta} = \vec{\eta}^\dagger M = 0$. This special property will allow us to construct N constants of motion, which we will call L_n, as

$$L_n \equiv \vec{\eta}^\dagger L^n \vec{\eta}. \tag{7.249}$$

To be concrete, we should prove that all L_n commute with the Hamiltonian, that they commute with each other and that they are independent (that is, that a set of L_n's being constants of motion does not imply that any other L_m is a constant of motion). Firstly, we need a general result regarding commutators that we will prove, namely, given A and B two linear operators, and n an integer,

$$[A, B^n] = \sum_{j=0}^{n-1} B^j[A, B]B^{n-1-j}. \tag{7.250}$$

We prove this by induction. Trivially, equation (7.250) is satisfied for $n = 1$. Assuming that equation (7.250) is valid for n, we have

$$\begin{aligned}[A, B^{n+1}] &= AB^{n+1} - B^{n+1}A = AB^nB - B^nBA = AB^nB - B^nAB + B^n[A, B] \\ &= [A, B^n]B + B^n[A, B] = \left[\sum_{j=0}^{n-1} B^j[A, B]B^{n-1-j}\right]B + B^n[A, B],\end{aligned} \tag{7.251}$$

from which equation (7.250) changing n for $n + 1$ trivially follows. We now apply equation (7.250) in the commutator $[H, L_n]$, with L_n given in equation (7.249),

$$[H, L_n] = \vec{\eta}^\dagger \sum_{j=0}^{n-1} (L^j [H, L] L^{n-j-1}) \vec{\eta}. \tag{7.252}$$

Inserting the quantum Lax equation, equation (7.229), into equation (7.252) above, we obtain

$$[H, L_n] = \vec{\eta}^\dagger \sum_{j=0}^{n-1} [L^j (AL - LA) L^{n-1-j}] \vec{\eta} = \vec{\eta}^\dagger [A, L^n] \vec{\eta} = 0, \tag{7.253}$$

where, in the last equality, we have used that $\vec{\eta}^\dagger A = A\vec{\eta} = 0$. The next point in our proof consists of showing that $[L_m, L_n] = 0$ for $n \neq m$. For this, we only need the following relation, given three linear operators A, B and C, which is trivial to prove,

$$[A, [B, C]] = [C, [A, B]] = [B, [C, A]]. \tag{7.254}$$

We set $A = H$, $B = L_m$ and $C = L_n$, with $m \neq n$, which immediately implies $[L_m, L_n] = 0$. Independence of the constants of motion is a little more involved, and we shall take a look at that in a minute. Before that, we show that L_1 and L_2 are familiar conservation laws. Indeed,

$$L_1 = \vec{\eta}^\dagger L \vec{\eta} = \sum_{j=1}^{N} p_j = P, \tag{7.255}$$

while L_2 is easily calculated as

$$L_2 = \vec{\eta}^\dagger L^2 \vec{\eta} = (\vec{\eta}^\dagger L) \cdot (L\vec{\eta}) = H. \tag{7.256}$$

Above, $\cdot$ is the usual scalar product in $\mathbb{R}^N$. We have just shown that L_1 is the total momentum, while L_2 is the Hamiltonian of the system. If $N = 2$, this is the end of it, so that, as we already knew, every two-body system is integrable. For more than two particles, this is no longer the case. This is what makes integrable systems non-trivial. A detailed study of what is actually conserved in the many-body problem involves the Bethe ansatz, which will be introduced in the second volume of this monograph.

The connection with the supersymmetric (product ground states) systems we studied earlier is now quite evident. We simply need to look at equation (7.185) where, to make a link to the above ansatz for the quantum Lax pair, we must replace its right-hand-side by a constant. Verifying that, in such case, we have integrable systems, is quite easy, and this is left as a problem to the reader.

7.3.3 Bosons and spinless fermions in one dimension

Here, we are going to solve exactly a couple of problems that are quite important in current research. In particular, we will obtain the exact three-body bound states of spinless bosons and fermions in one spatial dimension interacting with their respective LO EFTs. The observant reader may realize we have actually already solved these problems, and they are absolutely right. However, we feel it is good practice to solve at least some problems in several different ways (the observant reader may have realized that we have been doing this often throughout the book already), and we here solve them in the momentum representation using their respective Faddeev equations.

We begin with the three-boson problem. Hence, we need to solve the Skornyakov–Ter-Martirosyan equation, equation (7.37), in one spatial dimension. For $D = 1$, the T-matrix at energy $z = -|E| = -\hbar^2 Q^2/m < 0$ ($Q > 0$) is easily calculated as

$$\mathcal{T}_{\rm 1D}(z) = \frac{1}{\dfrac{1}{g_0} + \dfrac{m}{\hbar^2 Q}}. \tag{7.257}$$

For concreteness and book-keeping, we explicitly write the Skornyakov–Ter-Martirosyan equation in one dimension,

$$M(z;\,k) = 2\mathcal{T}_{\rm 1D}\left(z - \frac{3\hbar^2 k^2}{4m}\right)\int_{-\infty}^{\infty}\frac{dq}{2\pi}\frac{M(z;\,q)}{z - \epsilon(k) - \epsilon(q) - \epsilon(k+q)}. \tag{7.258}$$

To solve the equation above, we take inspiration in the one-dimensional version of the Bethe–Peierls boundary condition. At short-range, two particles approaching each other must have a relative wave function that behaves as

$$\psi(x) \propto a - |x|. \tag{7.259}$$

The Fourier transform of the absolute value distribution, $|x|$, is given, at large momenta $|k|$, by a term proportional to $1/k^2$. Therefore, we expect the wave function in the momentum representation to behave as $1/k^2$ when $|k|$ is large. For three particles, this is a bit more complex, but nonetheless the situation is similar. Hence, we propose a simple Ansatz of the form

$$M(z;\,k) = \frac{1}{1 + (k/b)^2}, \tag{7.260}$$

where b is some constant with dimensions of momentum, to be determined. The relation between M and the actual wave function is worked out in the problems. To show that equation (7.260) solves the Skornyakov–Ter-Martirosyan equation (7.258) with the correct value for b, which equals Q, we simply insert M into it. The only integral we will require is the following

$$\int_{-\infty}^{\infty}\frac{dq}{2\pi}\frac{1}{Q^2 + q^2 + kq + k^2}\frac{1}{1 + (q/Q)^2} = \frac{Q}{2(k^2 + Q^2)}\left[1 - \frac{Q}{2\sqrt{Q^2 + 3k^2/4}}\right]. \tag{7.261}$$

Inserting equation (7.261) above into equation (7.258), we obtain the solution $Q = 2/a$, so that the bound state energy is

$$E = -\frac{\hbar^2 Q^2}{m} = -\frac{4\hbar}{ma^2}. \tag{7.262}$$

We are going to solve now the three-fermion problem. This is slightly more involved, but still exactly solvable. The LO EFT interaction has the form

$$\langle k'|V_{\rm LO}|k\rangle = g_1 k'k, \tag{7.263}$$

where $g_1 = g_1(\Lambda)$ is the bare coupling constant. The T-matrix is given by

$$\frac{2\hbar^2}{m}\frac{k^2}{1/a + ik}, \tag{7.264}$$

where a is the odd-wave scattering length (see equation (5.24)).

The Faddeev equations for the three-fermion bound state with LO EFT interactions, equation (7.263), are obtained in a similar fashion as the Skornyakov–Ter-Martirosyan equation. We set once more, and without loss of generality, the total momentum to zero. In this case, it is straightforward to see that the matrix elements of the operators $M_i(z)$ have the form

$$\langle \vec{k}'|M_i(z)|\vec{k}\rangle = F_i(k_i')k'_{j\ell}. \tag{7.265}$$

Fermionic statistics (antisymmetry of the wave function) dictates that

$$F_i(k_j) = -F_j(k_i), \tag{7.266}$$

$$F_i(k_\ell) = F_\ell(k_i), \tag{7.267}$$

$$F_j(k_\ell) = -F_\ell(k_j). \tag{7.268}$$

Defining $F(k_i) = F_i(k_i)$, and making all momentum variables dummy, we easily arrive at the following integral equation

$$F(k) = 2\alpha\left(E - \frac{3\hbar^2k^2}{4m}\right)\int_{-\infty}^{\infty}\frac{dq}{2\pi}\mathcal{K}(k, q)F(q), \tag{7.269}$$

where α and the kernel are given by

$$\alpha(E) = -\frac{1}{mg_0/4\hbar^2 + \sqrt{m|E|/\hbar^2}/2}, \tag{7.270}$$

$$\mathcal{K}(k, q) = \frac{m}{\hbar^2}\frac{(k + q/2)(q + k/2)}{Q^2 + k^2 + kq + q^2}, \tag{7.271}$$

Above, we have defined the binding momentum Q via $E = -\hbar^2Q^2/m$, as usual. We can argue that, since $F(k)$ should behave as $1/k^2$ at large momenta, equation (7.273) has no non-trivial solution. This is because the integral

$$\int_{-\infty}^{\infty}\frac{dq}{2\pi}\mathcal{K}(k, q)\frac{1}{1 + (q/Q)^2} \to \text{constant}, \quad k \to \infty. \tag{7.272}$$

That is, for large k, the integral equation cannot be satisfied, since the left-hand-side of equation (7.269) behaves as $1/k^2$ while the right-hand-side necessarily behaves as $1/|k|$. This mismatch can be corrected in several ways. One way to do this is to work on a tight-binding lattice, which naturally regularizes the problem, and take the continuum limit. Another way to deal with this is by means of three-body interactions. Sekino and Nishida brilliantly resolved this problem in reference [19], using the dimer field trick. Here, we show their solution using standard first quantization.

The least singular (although not hyperspherically symmetric), non-trivial fermionic three-body interaction has the following form

$$V_3 = g_F k'_{ij} k_{ij} S(k_{i\ell} k_{j\ell}) + \text{Permutations}, \tag{7.273}$$

where $S(x)$ is the sign of x. The Faddeev equations turn into equation (7.269), albeit with the kernel given by that of Sekino and Nishida,

$$\mathcal{K}_{\text{S-N}}(k, q) = \frac{m}{\hbar^2}\left[\frac{(k + q/2)(q + k/2)}{Q^2 + k^2 + kq + q^2} - \gamma\right], \tag{7.274}$$

where γ appears analogously to what happened in the case of the three-body counterterm for the three-dimensional bosonic case. The way to calculate what (dimensionless) value γ should take is to cancel the constant limit of the asymptotic integral, equation (7.270). However, a good (and correct) guess can be obtained by reducing the kernel's large-q behaviour. Indeed, we may remove the term proportional to q^2 in the numerator of equation (7.273) by choosing $\gamma = 1/2$, obtaining

$$\mathcal{K}_{\text{S-N}}(k, q) = \frac{m}{4\hbar^2}\frac{3kq - 2Q^2}{Q^2 + k^2 + kq + q^2}. \tag{7.275}$$

It is now a simple yet tedious exercise to show that the integral equation (7.269), with the kernel replaced with equation (7.275), and $F(k) = 1/(1 + (q/Q)^2)$ solves the integral equation if $Q = 2/a$.

Yet another very clean method to solve the problem is to use Danilov–Pricoupenko filtering. This is left for an interesting problem below.

Problems

1. Write down the Faddeev equations (7.17) for the three Faddeev components T_1, T_2 and T_3 in matrix form, considering $(T_1, T_2, T_3)^{\mathrm{T}}$ as a vector, when no three-body interaction is present.
2. Prove that the energy solutions to equation (7.67) are given by equation (7.68).
3. Consider three bosons in the supersymmetric formalism, with superpotentials W_i of the form of equation (7.182). Assume that

$$W^{(2)}(x) = c_2\,\text{sgn}(x),$$
$$W^{(3)}(x) = c_3\,\text{sgn}(x).$$

Obtain the conditions on c_2 and c_3 for the state annihilated by A_i to be a physically admissible ground state, and calculate it.

4. Calculate, when possible, the ground state and ground state energy of non-relativistic three-body systems in one dimension with the two-body interaction potentials

$$V_1(x) = \frac{A}{x^2},$$
$$V_2(x) = B\delta(x),$$
$$V_3(x) = Cx^2,$$

where x is the relative coordinate between any two particles. To do this, consider the superpotentials in equations (7.186)–(7.188).

5. Show that the momentum representation of a three-body bound state with vanishing total momentum and energy E is given by

$$\psi(\mathbf{k}_1, \mathbf{k}_2) = \frac{M(E; \mathbf{k}_1) + M(E; \mathbf{k}_2) + M(E; -\mathbf{k}_1 - \mathbf{k}_2)}{E - \epsilon(\mathbf{k}_1) - \epsilon(\mathbf{k}_2) - \epsilon(\mathbf{k}_3)}.$$

6. Use Danilov–Pricoupenko filtering for the fermionic three-body problem in one dimension. That is, impose that $F(k_*) = 0$ for some momentum k_* and obtain the resulting integral equation. Then take the limit $k_* \to \infty$ to show that $F(q) = 1/(1 + (q/Q)^2)$ is a solution with $Q = 2/a$. This is easiest to show by proving that the resulting kernel equals Sekino–Nishida kernel, equation (7.275).
7. Obtain the next-to-leading-order correction to the ground state energy of three bosons within the strong-coupling expansion to solve Mattis equation, equation (7.130).
8. Show that the three-body problems with pairwise interaction potentials given in problem 4 are indeed integrable.

References

[1] Joachain C J 1975 *Quantum Collision Theory* (Amsterdam: North-Holland)

[2] Gogolin A O, Mora C and Egger R 2008 Analytical solution of the bosonic three-body problem *Phys. Rev. Lett.* **100** 140404

[3] Gradshteyn I S and Ryzhik I M 2007 Tables of Integrals *Series and Products* (New York: Academic)

[4] Danilov G 1961 On the few-body problem with short-range forces *Sov. Phys. - JETP* **13** 349

[5] Weiss R, Bazak B and Barnea N 2015 Nuclear neutron-proton contact and the photo-absorption cross section *Phys. Rev. Lett.* **114** 012501

[6] Yamashita M T, Rosa D S and Sandoval J H 2018 Few-body techniques using momentum space for bound and continuum states *Few-Body Syst.* **59** 19

[7] Pricoupenko L 2010 Crossover in the Efimov spectrum *Phys. Rev.* **A82** 043633

[8] Bedaque P F, Hammer H W and van Kolck U 1999 Renormalization of the three-body system with short-range interactions *Phys. Rev. Lett.* **82** 463

[9] Faddeev L 1961 Scattering theory for a three-particle system *Sov. Phys. - JETP* **39** 1459
[10] Glöcke W 1983 *The Quantum Mechanical Few-Body Problem* (Berlin: Springer)
[11] Hammer H W, Nogga A and Schwenk A 2013 Colloquium: three-body forces: from cold atoms to nuclei *Rev. Mod. Phys.* **85** 197
[12] Beane S R 2010 Ground-state energy of the interacting Bose gas in two dimensions: an explicit construction *Phys. Rev.* A **82** 063610
[13] Valiente M 2019 Three-body repulsive forces among identical bosons in one dimension *Phys. Rev.* A **100** 013614
[14] Valiente Cifuentes M 2010 Few quantum particles on one dimensional lattices *Scattering, binding and transport. PhD Thesis* (Berlin: Humboldt-Universität zu Berlin)
[15] Mattis D C 1986 The few-body problem on a lattice *Rev. Mod. Phys.* **58** 361
[16] Petrov D S 2012 The few-atom problem *Many-Body Physics with Ultra-Cold Gases* Lecture Notes of the 2010 Les Houches Summer School
[17] Cooper F, Khare A and Sukhatme U 1995 Supersymmetry and quantum mechanics *Phys. Rep.* **251** 267
[18] Calogero F 2001 *Classical Many-Body Problems Amenable to Exact Treatments* (Berlin: Springer)
[19] Sekino Y and Nishida Y 2021 Field-theoretical aspects of one-dimensional Bose and Fermi gases with contact interactions *Phys. Rev.* **A103** 043307

IOP Publishing

Strongly Interacting Quantum Systems, Volume 1

Few-body physics

Manuel Valiente and Nikolaj Thomas Zinner

Appendix A

Useful formulae

A.1 Poisson's summation formula

Let $f(x)$ be a complex function of a real variable, and $\tilde{f}(k)$ its Fourier transform, defined as

$$\tilde{f}(k) = \int_{-\infty}^{\infty} f(x)e^{-2\pi ikx}.$$

Then if for some $\epsilon > 0$ the following condition is satisfied

$$|f(x)| + |\tilde{f}(x)| \leqslant C\frac{1}{(1 + |x|)^{1+\epsilon}}, \tag{A.1}$$

for some real positive constant C, the following holds

$$\sum_{n=-\infty}^{\infty} f(n) = \sum_{k=-\infty}^{\infty} \tilde{f}(k). \tag{A.2}$$

The above expression is called Poisson's summation formula.

Throughout the book, equation (A.2) is mostly written slightly differently. For instance, we are usually interested in functions of the form $F(k)$, with $k = 2\pi n/L$, where n is an integer. Therefore, equation (A.2) is rewritten as

$$\sum_{k} F(k) \equiv \sum_{n=-\infty}^{\infty} F\left(\frac{2\pi n}{L}\right). \tag{A.3}$$

The function f in Poisson's summation formula is $f(n) = F(2\pi n/L)$ and, therefore, its Fourier transform $\tilde{f}(m)$ is given by

$$\tilde{f}(m) = \frac{L}{2\pi}\int_{-\infty}^{\infty} dqF(q)e^{-imqL}. \tag{A.4}$$

doi:10.1088/978-0-7503-3087-9ch8

Putting all this together, we obtain for the Poisson's summation formula, in the most convenient form for our purposes,

$$\frac{1}{L}\sum_{k}F(k) = \int_{-\infty}^{\infty}\frac{dq}{2\pi}F(q) + \sum_{m\neq 0}\int_{-\infty}^{\infty}\frac{dq}{2\pi}F(q)e^{-imqL}. \tag{A.5}$$

The formula above is also valid in higher dimensions, where we replace k with $\mathbf{k}$, m with $\mathbf{m}$, L with L^D, 2π with $(2\pi)^D$, and the integral over dq with an integral over $d\mathbf{q}$ in $\mathbb{R}^D$, where D is the spatial dimension.

A.2 Integrals in effective field theory

When solving many of the problems using low-energy interactions, i.e., effective field theory (EFT), we encounter a number of typical integrals. In three dimensions, these are given by

$$J_n(z) = \int \frac{d^3q}{(2\pi)^3}\frac{q^n}{z - \dfrac{\hbar^2q^2}{m}}, \tag{A.6}$$

with $n = 0, 2, 4\ldots$. The values of z we are mostly interested in are $z = \hbar^2k^2/m + i\eta$, with $\eta \to 0^+$, that is, on the energy shell. In three dimensions, all the integrals in equation (A.6) are divergent. Therefore, we calculate them using a momentum cutoff Λ, such that

$$J_n(z) = \frac{m}{\hbar^2}\int_0^{\Lambda}\frac{dq}{2\pi^2}\frac{q^{n+2}}{k^2 - q^2 + i\eta}. \tag{A.7}$$

To obtain the above quantities (with $z = \hbar^2k^2/m + i\eta$), we first separate real and imaginary parts as

$$J_n(z) = \frac{m}{\hbar^2}\left[\int_0^{\Lambda}\frac{dq}{2\pi^2}\frac{q^{n+2}(k^2 - q^2)}{(k^2 - q^2)^2 + \eta^2} - i\int_0^{\Lambda}\frac{dq}{2\pi^2}\frac{q^{n+2}\eta}{(k^2 - q^2)^2 + \eta^2}\right]. \tag{A.8}$$

We use Plemelj's formula, which states that

$$\mathrm{P}\frac{1}{k^2 - q^2} = \lim_{\eta\to 0}\frac{k^2 - q^2}{(k^2 - q^2)^2 + \eta^2}, \tag{A.9}$$

where P denotes principal value, and

$$\pi\delta(k^2 - q^2) = \lim_{\eta\to 0^+}\frac{\eta}{(k^2 - q^2)^2 + \eta^2}. \tag{A.10}$$

The above two relations, when inserted into equation (A.8), yield

$$J_n(z) = \frac{m}{\hbar^2}\left[\mathrm{P}\int_0^{\Lambda}\frac{dq}{2\pi^2}\frac{q^{n+2}}{k^2 - q^2} - i\int_0^{\Lambda}\frac{dq}{2\pi}q^{n+2}\delta(k^2 - q^2)\right]. \tag{A.11}$$

Using the properties of the Dirac delta function, the imaginary parts of $J_n(z)$ are straightforwardly given by ($k > 0$)

$$\mathrm{Im}\,(J_n(z)) = -\frac{mk^{n+1}}{4\pi\hbar^2}. \tag{A.12}$$

For the real parts, we first calculate $J_0(z)$, to $O(1)$, that is, discarding inverse powers of the cutoff, and then extract $J_n(z)$, for $n > 0$, using a recurrence relation that we obtain below. We have

$$\begin{aligned} \mathrm{Re}\,(J_0(z)) &= \frac{m}{2\pi^2\hbar^2}\mathrm{P}\int_0^\Lambda \frac{q^2}{k^2 - q^2} \\ &= \frac{m}{2\pi^2\hbar^2}\mathrm{P}\int_0^\Lambda \frac{q^2 - k^2}{k^2 - q^2} + \frac{m}{2\pi^2\hbar^2}\mathrm{P}\int_0^\Lambda \frac{k^2}{k^2 - q^2} \\ &= -\frac{m\Lambda}{2\pi^2\hbar^2} + O(\Lambda^{-1}). \end{aligned} \tag{A.13}$$

The recurrence relation is easily obtained by noticing that, for $n \geqslant 1$,

$$\frac{q^{n+2}}{k^2 - q^2} = \frac{k^2 q^n}{k^2 - q^2} - q^n. \tag{A.14}$$

Therefore, for even $n \geqslant 2$, we have

$$\mathrm{Re}\,(J_n(z)) = k^2\,\mathrm{Re}\,(J_{n-2}(z)) - \frac{m}{2\pi^2\hbar^2}\frac{\Lambda^{n+1}}{n + 1}. \tag{A.15}$$

Milton Keynes UK
Ingram Content Group UK Ltd.
UKHW051501031023
429866UK00002B/13